SOUTHER

· TIMES ·

Contents

The next issue of Southern Times, No 9, will contain:
Farnborough Air Show traffic
Stored locos at Fratton, Part 2 of Ashord to Dover,
71 more ‘Leaders’...?
The Elham Valley by A Earle Edwards
Cycling excursion in the 1930s, a pictorial ‘Withered Arm’
More early colour - S C Townroe - From the Footplate etc etc.

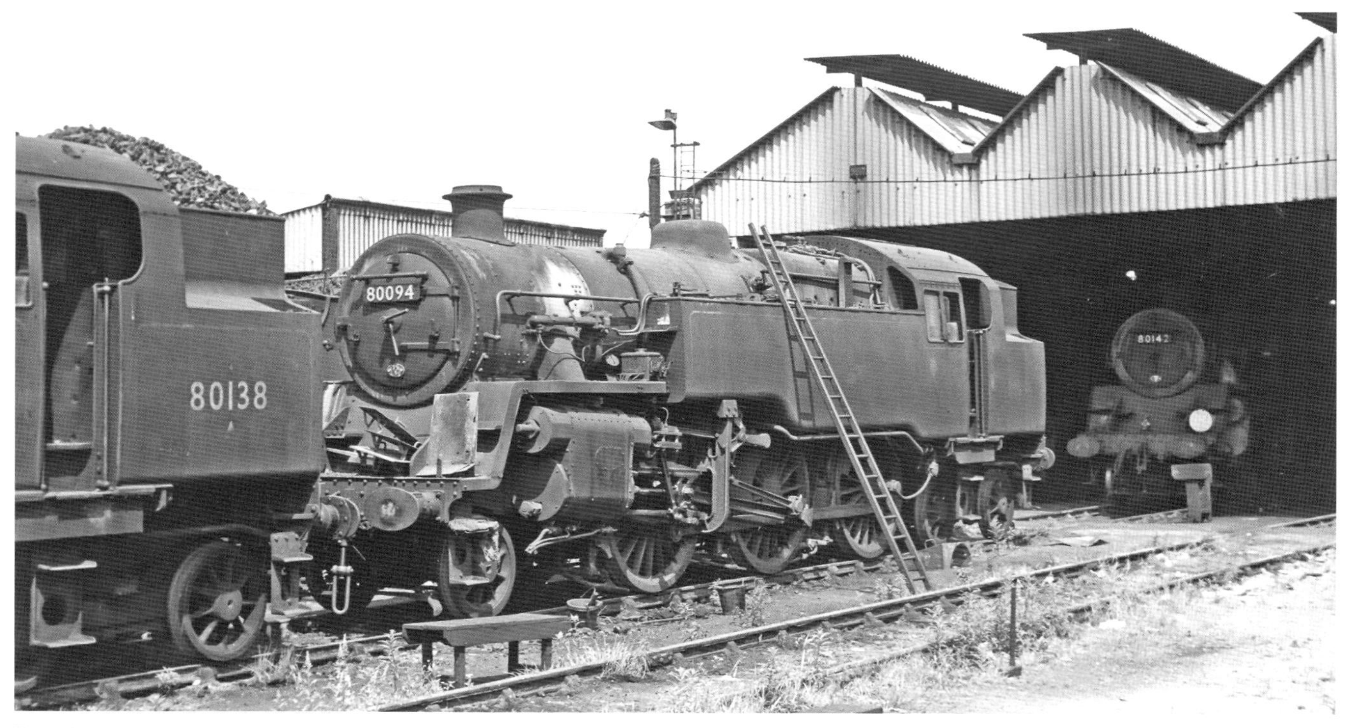

Frontispiece: And speaking of trios of tank engines, here we have Nos 80138, 80094 and 80142 of the BR4 type at Redhill. Externally at least, the first two referred to are in less than pristine condition whilst look also at the coal in the bunker of No 80138 which appears none too secure. The front framing of No 80094 appears to contain several of the screens associated with the self cleaning fittings from within the smokebox whilst the presence of the ladder would all add to the conclusion this engine at least was receiving attention of some sort. *Transport Treasury*

Front Cover: Variations upon a theme at Waterloo. Left to right: units 3135, 3106, 4391, and an unidentified suburban unit. Nos 3135 and 3106 are pure SR designs, No 3135 dates from 1938 and electrification of the Mid-Sussex line, No 3106 from the year earlier and the electrification of the Portsmouth line through Guildford. As we have said before, green so suited the sets, BR blue immediately giving a dated appearance. All these sets had gone by 1972. Alongside No 4391 is a Sub unit, once a familiar sight on the lines in and out of all the SR termini. *Les Price*

Rear Cover: Surely a rarely photographed occasion: three 'Z' class 0-8-0Ts alongside Exeter (St Davids) Middle signal box (Red Cow Crossing) and all waiting their next duty banking trains to Exeter Central. No date, but certainly post 1957 as locos two and three, Nos 30954 and 30955, both display the later BR crest. It is doubtful if all three would be used to bank a single train unless of course it was to be used as a means of returning one or more to the SR proper. As we know, sadly none from this small class of just eight engines was preserved. *MHW / Transport Treasury*

Copies of many of the images within **SOUTHERN TIMES** are available for purchase / download.
In addition the Transport Treasury Archive contains tens of thousands of other UK, Irish
and some European railway photographs.

ISBN 978-1-913251-75-8

First Published in 2024 by Transport Treasury Publishing Ltd.,
16 Highworth Close, High Wycombe, HP13 7PJ

admin@ttpublishing.co.uk or for editorial issues and contributions email to **southerntimes@email.com**

Printed in the Malta by the Gutenberg Press.

INTRODUCTION

The railways appear to been afflicted with a number of wash-outs in recent times – writing this in March 2024. In part these are due to a number of unseasonably heavy deluges but it must also beg the question over the age of what is sometimes a creaking infrastructure. In similar vein the recent (well March 2024) derailment of an early morning service near Woking which hit an obstruction in the dark at 85mph left by overnight p/way work does ask the question, are we really learning from past experience?

I am told by an 'insider' that drones and the like are now used to check for track and infrastructure defects and whilst the use of modern technology is to be applauded have we really bettered the era when there was a man who walked his length daily aware of the areas of concern and who could monitor them accordingly? No, I am not getting specifically nostalgic, just concerned, for its seems we have to reinvent the wheel every so often and the phrase, 'We will learn and put measures in place' is starting to wear a little thin. (I could get on my high horse at this point but I will refrain from doing so, for the same phrase almost seems to be a cop-out for much in the way of business and goverment nowadays.)

Back to history, I am delighted to report that the Transport Treasury archive has been buoyed recently with the acquisition of the colour collection from the late R C Riley. There can be few who have have not heard or likely seen examples of 'Dick' Riley's material and whilst an amount may have been published in the past I can promise there is plenty more to come; watch this space for more detail.

Indeed one of the more onerous (?) tasks I have been undertaking in relation to this collection is to undertake an initial sort into 'yes', 'no' and 'themed' ideas - well someone has to do it...! Ignoring specifics one thing came through at the outset which was simply that I was looking at images portraying what is almost a different world. Yes I know we might say the same about most of the photographs in Southern Times, but in this particular case it was the material in the background that so enhanced the railway subject it was almost like looking at a genuine snapshot in time.

Let me continue on this theme by stating that over the years I have had the privilage of viewing what is probably tens of thousands of railway photographs and it could be all too easy to even become bored with the subject. Clearly my 'illness' relating to railways must be somehwat deeply ingrained for I still get a kick from looking at new images, perhaps maturity also meaning that it is now not just views of pet or specific interests that I get a kick from but railway images generally. Consequently the subject may be Wareham or Wick, Hastings or Haverfordwest; I suspect most of us are the same - we just like trains and railways.

Some of the time taken viewing this new material was during a period of illness at the start of 2024. To all those who kindly enquired as to my welfare and passed on good wishes, a sincere thank you. I might have had my share of the NHS over the past few months but I am now hopefully able to resume work and indeed intend to continue for some time to come.

With new historic photographic material coming to light it is also a pleasure to record that new facts and figures can also appear; information on the Loco Building Programme for example as appears in this issue and likewise the statistical detail referred to by Tony Hillman in 'Treasures from the Bluebell Museum'.

All this fills me with confidence that what we are doing – not just me but my fellow editors at Transport Treasury – and hopefully getting it right. We intend to continue for a long time to come.

And just to say we have deliberately not forgotten our promised special on the 'Tavern Cars'. As I write this it is basically complete and just needs me to get to the offcvie to answer a couple of queries before it goes off to print. As they say, 'watch this space'!

Kevin Robertson

Frontispiece - opposite: (page 3) – And speaking of trios of tank engines, here we have Nos 80138, 80094 and 80142 of the BR4 type at Redhill. Externally at least, the first two referred to are in less than pristine condition whilst look also at the coal in the bunker of No 81038 which appears none too secure. The front framing of No 80094 appears to contain several of the screens associated with the self cleaning fittings from within the smokebox whilst the presence of the ladder would all add to the conclusion this engine at least was receiving attention of some sort. ***Transport Treasury***

GENUINE early colour, not a modern day recolouring. Whilst in the process of compiling this issue Robin at TT announced he had come across a small batch of Duffay colour views mostly it appears from the 1940s – but we not have actual dates. Several Southern examples, one of which will appear on the cover of the next issue with others inside. In addition there Transport Treasurywas this superb showing of No 2329 Stephenson at Waterloo, Now this is how an N15X should look, the single disc indicating a Waterloo to Southampton Terminus service. *Transport Treasury*

N15 X-'tra'

As mentioned in the Introduction, for a period at the start of 2024 I was temporarily incapacitated and in consequence restricted as to movement. Unable to undertake much if any work during this period, it was during this sojourn that I picked up loose copies of various old magazines as a welcome change from radio / television. One of these was *Trains Illustrated* No 15 (October / November 1949) and where I chanced upon an article by O J Morris on the N15X. Running over three sides for the majority it refers to the history of the type - as we ourselves featured in Issue 7.

In addition to the known facts, Mr Morris gives some additional information which does not appear to have perpetuated elsewhere. Firstly we are told that No 333 *Remembrance* ran for two years in grey, but not just a potentially short lived photographic grey, for instead this was '...thoroughly well done...' and included lining and lettering to the same standard. Morris adds, '... and aroused much interest for the dignified blend of sympathy and decorum'.

Shortly after he refers to the debate over the naming, or otherwise of the class at the time of their conversion into tender engines. 'When the engine (No 333) came to be converted to an N15X, it was touch and go whether these plaques would go into the melting pot, and the writer, always a collector of old engine brasses, had already put in an order to the Stores Dept. for their acquisition under the heading of 'old material'. Fortunately or otherwise, according to viewpoint, fresh counsels prevailed.'

In the opinion of the present writer, by far the most interesting piece of new information on the type concerns their removal from Brighton, as tank engines, to Eastleigh ready for conversion into tender type. Morris admits that even allowing for the generous Brighton loading gauge, the sheer size of the 'L' class as tank engines had been a nuisance from the start with the arch entrances to Brighton loco having cut back to allow the cab eaves to pass. Movement to Eastleigh was similarly problematical owing to one particular bridge (not identified) near to Fareham though which

No 331 in original form recorded in Clapham Common and with signs of the Brighton overhead still present. The engine is hauling Maunsell stock and it will be noted dwarfs the height of the train. ***Transport Treasury***

The same engine as photographed on the previous page, but now No 2331 and carrying the name *Beattie*, running light out of Waterloo having probably brought a service from Basingstoke. Basingstoke – Waterloo semi-fast workings were the main duties for the class for many years. *David H Bayes / Transport Treasury*

the cab declined to pass. We might ask was this even established though experience?

Accordingly when a member of the 'L' type was to be sent from Brighton to Eastleigh, the latter would remove the upper part of the cab and both front and rear weatherboards, these parts then stacked and lashed on top of the coal bunker. The engine was then worked in steam, but presumably 'light' to Eastleigh with the crew exposed to whatever the elements were on the occasion. Morris comments, '...the bunker looking something like a junk stall. I believe that modern locomotion has never presented a more fantastic picture to less responsive cameras.' Unfortunately no image of such a move has been located.

The rest of the piece refers to the changes wrought by the actual conversion which we have already covered, although Morris adds his own name to that of others mentioned when it came to the naming of the class.
A final comment is one which should be considered after the reader has digested the next article in this issue. Considering the lack of a large tank engine on the Southern, not withstanding the SR's Chief Engineer being so opposed to large tank engines, might there have been consideration to modifying the original L into a machine with a greater route availability?

The 1948 Southern Region Locomotive Building Programme

The intentions of the Southern Railway at the time of nationalisation are well known; electrification moving westwards with other services potentially to be worked by diesel-electric locomotives. Steam was very much 'yesterday's traction' although it was recognised to achieve this aim would have taken considerable investment.

As a private company, had nationalisation not occurred, then history tells us Bournemouth might well have been reached by third-rail in 1955, other locations, Salisbury for example and perhaps beyond as well as the countless feeder, cross-country and other connecting routes not individually defined although some of these may well have been given over to diesel rather than be 'juiced'. (Should anyone have details of the proposals covering every inter-connecting line we would be very pleased to receive details.)

Returning to 1946 however, and we know the senior officers met on a regular basis to approve new works, as well as new construction, the latter covering both locomotive and rolling stock. In the case of locomotives and rolling stock the SR were hampered in exactly the same way as the other railways at the time, shortage of materials and similarly shortage of skilled men. It was also their task to attempt to foresee potential traffic patterns and in consequence needs into the future whilst at the same time continuing to address the backlog in repairs and maintenance as well as making good deficiencies in locomotives and rolling stock in consequence of WW2. 1948 was not an easy time to be a railway manager.

At this point then we might introduce a file held at the National Archives Rail 1188/207. Notwithstanding having spent much time at Kew, the NRM and other archives in my earlier pursuit of information on 'Leader' this revealed it self to be a new source of information and well worth including in ST not just for the additional information in reveals on Mr Bulleid's work, but also some hitherto likely unknown aspect of what might

Taken from the cover of the May 1948 issue of the *Southern Region Magazine*, Fairburn 2-6-4T No 42198 is leaving Waterloo with the 10.54am Waterloo – Salisbury service. The caption accompanying the image commented that such services would normally be in the hands of a 'U' or 'N15X' locomotives, the 2-6-4T having a slight increase in tractive effort over the other two. 'First reports are good'.

Q1 No 33005 at Fratton on 24 July 1958. Forty of the class were built and if Bulleid had had his way there would have been another 11 in 1948. The tenders of these engines were always 'lively' especially when low on water and in consequence 2½ inches of concrete to the interior base of some – possibly all – the tenders. Trials were also run with at least one member of the class, No C36 attached to a West Country tender especially painted black and with the driving controls duplicated on the opposite side of the backhead for tender first running. ***Brian Hilton / Transport Treasury***

have been seen on the Southern Region in the years immediately post nationalisation.

We start with a memorandum (sender and recipient not confirmed but believed to have been Elliott at Waterloo and Riddles and Marylebone respectively) dated just one year after nationalisation, on 29 January 1949.

'Mr Bulleid and Mr Marsh called to see me this morning. Mr Bulleid stated that it had been agreed that the British railways should go forward with the building of types of locomotive, for which designs have already been got out, during the years, 1948, 1949 and possibly into 1950.

'The C.R.O. has approved a figure which Mr Bulleid has sent forward to the Railway Executive, stating that we have yet to build 72 tank locomotives. As is well known, five engines of the 'Leader' class are already being built, and provided these prove satisfactory on test, there are the remaining 20 engines of this class to build.

This leaves 47 tank engines (72 - 5+20), and Mr Bulleid makes the strong suggestion that we should go forward with the building of 47 additional 'Leader' class tank engines, provided of course they are satisfactory, in which case our compliment of tank engines would be of our own design, and we should not have to accept 47 tank engines already in existence from another Region. Mr Bulleid stressed the importance of this, in view of the fact that electrification would appear to be receding, and that we have to provide tank engines for the whole region and not for a limited portion of it.

'It may be mentioned that our total compliment (steam?) at the end of 1947 was 690, and therefore 72 big tanks would be a very valuable betterment to the tank fleet.' Revelations indeed. At long last a recognition that the SR was indeed desperate for a large tank engine, of the type previous shunned following the Sevenoaks disaster of two decades earlier. We shall continue with the theme of 'imported' large tank engines shortly but the present it is also worthwhile pointing out that how Bulleid not only attempts to convince all others that his Leader design will be a success – the fact it was still 17 months before the first (and only one) to steam began its trails from Brighton – but that the number of engines

he proposed to build to the design has suddenly jumped to a potential 72 (36001- 36072…). Bulleid's reputation had however preceded him here and wise council by others was also urging caution until the new design had proven itself; as we of course know, it never did.
For a newly nationalised industry things now move fast, for less than two weeks later on 11 February, Robin Riddles in charge of the Locomotive and Rolling Stock Building Programme for the whole of British Railways writes back to Elliott at Waterloo from his office at Marylebone.

We need not include the letter in full, suffice to say Riddles is aware the Leader class as being a 'special double-bogie type'. He then adds , 'In particular I notice that five mixed traffic steam tank locomotives on the 1947 building programme are expected to be completed at Brighton this year. These engines are of a special double-bogie type which Bulleid has designed. There are 31 additional mixed traffic tank engines on the 1948 programme all of which are estimated to be completed in 1949. 20 of these, I understand, are also of the special type.'

'Whilst I am at all times anxious to encourage new developments, I am sure you will agree with me that the history of unusual locomotives has not always been particularly happy'. Riddles does not elaborate on this and so we cannot be clear if he was being polite with reference to the Bulleid Pacific type, or it was instead a more general statement.

He continues, 'In the circumstances therefore, I feel that it would be wise to limit the construction of these special engines to the first five to be built this year. I propose that these five locomotives should run in traffic sufficiently long to enable experience to be gained of their performance before proceeding further with any further engines of this type. I should therefore like your agreement to the 20 programmed to be built in 1949 being postponed.'

Riddles was also clearly aware of the desperate need for large tank engines on the Southern and makes the suggestion, '…if possible, that postponement of the 20 special tank engines in 1949 need not entail the construction of 20 engines of alternative design for the Southern Region. I could fill the locomotive building capacity available at Brighton …..by engines on the London midland Region 1949 programme which at present are planned to be carried over into 1950.' Riddles concludes by enquiring of Elliott as to the proposed power capacity required for the mixed traffic engines desired.

At this point we should attempt to bring into play the

Fairburn 2-6-4T No 42098 at Dormans on 11 September 1950. This was one of the Brighton built members of the class. ***R C Riley / Transport Treasury***

relationship that existed between Bulleid, Elliott, Riddles and to an extent, Missenden. Bulleid as we know had previously enjoyed an excellent relationship with Missenden when the latter was General Manager at Waterloo. In 1948 though, Missenden had been elevated to the post of Chairman of the Railway Executive covering all four of what were now the 'regions'. From other study it appears it was not a particularly popular appointment either, the GM of the smallest of the 'Big Four' now having control over all its previous larger neighbours. With Elliott now in charge at Waterloo no such similar 'cosy' relationship was possible. Bulleid remained as CME of the Southern Region but was now under the overall control of Riddles, similarly Elliott, even if he had wanted to, no longer had the authority of his predecessor. The whole was a logistical nightmare, and whilst the new British Railways was not yet in a position to dictate all policy it was clear Riddles was adopting a cautious approach.

A few short weeks later on 19 March 1948 we learn Bulleid was evidently being pressured by Riddles' assistant at Marylebone, R C Bond, to accept a large tank engine of existing LMS/ LNER type; this in place of an additional 20 'Leader' type engines. We might even ask if pressure from the SR Traffic and Motive Power departments might well have played a part as certainly both were in desperate need for this type of machine.

Bulleid however, and not surprisingly was sticking to his guns quoting first the off stated note that his Leader design would be able to run over the vast majority of SR lines. He consequently dismisses the LNER 2-64T as it '...can only run over 'Merchant Navy' routes.

'The LMS 2-6-2T is quite inadequate in capacity except for use on the Isle of Wight. It is the equivalent to the Central Section D3 confined to pull and push work. It would not replace the M7.

'The LMS 2-6-4T is an engine similar to the 'Rivers' of 1927. It is too wide across the cylinders to be used on the Hastings route. In can operate over 'N' and 'U' routes only. It is not powerful enough to handle the loads specified at the speeds required. It carried too little coal and water. '

Bulleid goes on to criticise the 2-6-4T comparing it unfavourably his Leader design, he also makes the (unfair) comparison of stating the 2-6-4T could also not take its place in a link of 'West Country' engines but then surely the two would never be likely to perform the same duties. Interestingly no mention is made of the possible import of any Swindon design, the 61xx the obvious

The only sign of compromise came when Bulleid added

LNER L1 No 67711, another 2-6-4T the design of which was once considered as a possible contender for a large tank engine suitable for the Southern Region. The engine is seen here on its home turf at Wickham Market Junction. Folklore has it that these engines had a habit of rattling so much that pieces would reguarly work loose. ***Dr I C Allan / Transport Treasury***

BR Standard class 4 No 80033 on a Brighton to Tonbridge service, 24 April 1962. Of the 155 engines in the class no less than 130 were built at Brighton, this particular example taking to the rails in Sussex in March 1952. So far as the Southern Region was concerned they would eventually spread throughout the region being seen on everything from passenger to freight, shunting and permanent way duties – realistically exactly the type of tank engine the SR had needed for years. *Transport Treasury*

'....the CCE's acceptance of a leading truck in front of six-coupled wheels is essential. He has opposed such an arrangement since the Sevenoaks accident'.

The CCE added that from his perspective 31 mixed traffic engines, 20 x West Country, and 11 Q1 would be acceptable – but acceptable to whom? To Bulleid, to the Traffic / Motive Power Departments, we might suspect the first named only, unless of course the Q1 was to be developed with a better tender and improved rear lookout and the controls duplicated for rearwards running as indeed had been briefly experimented upon a short time earlier.

Before moving on we might also refer back to the mention of 2-6-2T locos for the Isle of Wight. Likely this was the first time such mention had been made but it would of course come up for discussion again 15 years later when BR were considering a steam replacement for the O2's. Finaly we should mention that Mr Bulleid's comparisons with the LMS 2-6-4T and 2-6-2T classes vis-a-vis the D3 and M7 is really rather unfair. Both the LMS types were far more modern than the two Southern classes and surely would have performed better. The whole gives the impression Bulleid was simply not prepared to look beyond his own stock and his own beliefs.

A final decision appears to have been made around 21 May and seemingly with Bulleid either sidelined or simply outmanoeuvred. It comes in the form of a memorandum, we think from Elliott to Riddles, in which Elliott refers to satisfactory trials having previously taken place with two LMS Class 4MT tank engines. 'I am pleased to say that the LMS Class 4 tank engine can be accepted for work in this Region. It is a modern and efficient locomotive, well suited to our requirements in many respects. As you know, our present fleet of tank engines is the oldest in the country, and we have hardly any under 30 years old, therefore, it is a machine for which we can find much work on secondary trains. You will of course appreciate, that it cannot be a substitute for a tank engine of the capacity which Bulleid has designed (Leader class),..........the 4P cannot meet all of these needs, but without doubt, will meet some of them.'

What follows is as they say history. No further Q1s were built and instead it was LMR class 4MT tanks and BR standard Class 4 tank engines that dominated the Southern Region for some years. A total of 41 Fairburn type LMS 2-6-4T locos were built at Brighton in 1950/51. Allocated running numbers 42066-42106, as they were built for SR use the original intention had been to number them in the 38xxx series.

They were followed by batches of the BR Standard class 4 tank, which eventually came to be seen all over the Southern, as such the LMS type were considered non-standard and all had been transferred away by the end of 1959.

This though is not the end of file Rail 1188/207, for what followed on is a raft of correspondence on No 36001 and her sisters and how BR felt best to respond to outside queries about the new design. It mainly covers the period 1949/1950 with one further item from as late as 1958 and goes to show how BR attempted to manage what was a 'difficult' problem form a PR perspective. This will feature in Issue 9 of 'ST'.

Last Train from Allhallows

Howard Cook

WITHDRAWAL OF PASSENGER TRAIN SERVICES BETWEEN GRAVESEND-ALLHALLOWS-GRAIN and CLOSURE OF ALLHALLOWS STATION TO FREIGHT TRAIN TRAFFIC

On and from MONDAY, 4th DECEMBER, 1961 passenger train services between GRAVESEND CENTRAL-ALLHALLOWS-GRAIN WILL BE WITHDRAWN and the following stations and halts CLOSED to passengers:-

ALLHALLOWS-ON-SEA
BELUNCLE HALT
CLIFFE
DENTON HALT
GRAIN
HIGH HALSTOW HALT
MIDDLE STOKE HALT
SHARNAL STREET
STOKE JUNCTION HALT
URALITE HALT

Additional 'buses serving the area will be provided for business and school purposes.

ALLHALLOWS-ON-SEA station WILL BE CLOSED to freight train traffic.

Collection and delivery services for parcels and freight sundries together with facilities for full truck load traffic at other stations and sidings in the area will remain.

Further information may be obtained from the Station Master at CLIFFE (Telephone No. 235) or GRAVESEND (Telephone No. 379) or from the Line Traffic Manager, Southern Region, British Railways, South Eastern Division, 61 Queen Street, London, E.C.4 (Telephone WATerloo 5151, Extension 227).

Enquiries concerning 'bus services in the area should be addressed to:-

THE MAIDSTONE & DISTRICT MOTOR SERVICES LTD.
KNIGHTRIDER HOUSE, MAIDSTONE, Tel. No. 2211
46 OVERCLIFFE, GRAVESEND, Tel. No. 1316

SOUTHERN

Closure notice for Allhallows and Grain branch services as posted at Stoke Junction Halt 2 December 1961. *A E Bennett 5884A / Transport Treasury*

'At 8.38p.m. on Sunday, 63-year old Robert Solly will set the signal at Allhallows station to closed as the Flyer departs for the last time.' So ran an article in the *Gravesend Reporter* dated 2 December 1961. The train that had transported crowned heads of Great Britain and Eastern royalty was not expected to run empty as demand for tickets had been high and it was likely that the Flyer would have one of its busiest journeys ever.
Mr Solly had been at Allhallows station doing the same job since it opened in May 1932. Officially described as a signalman, he was also ticket office clerk, 'clipper', collector, porter, cleaner and odd-job man. Whilst being the sole operator on duty in the platform signal box, he was rarely alone as the train crews would drop in for a cup of tea which he brewed almost continually on an open stove.

The short 1¾ mile-long branch line from Stoke Junction, situated between Middle Stoke and Grain Crossing Halts on the Hundred of Hoo Railway, opened partly by the Southern Railway on 14th May 1932 and fully, two days later. Following the west side of Yantlett Creek, separating the Isle of Grain from the mainland, the line was optimistically doubled in 1935 but singled again in 1957. The lengthy, canopied island platform of standard concrete components was pre-fabricated at Exmouth Junction Works and both faces had run-round loops.

The first train to run was a special train originating in London headed by an ex-SER 0-6-0T Class R locomotive, designed by James Stirling, conveying 700 day trippers to attend the ceremonial opening. 11 members of this class of 25 were later re-built with Wainwright-design boilers, 2 surviving until shortly before the closure of the branch but by then re-deployed elsewhere.

The Southern Railway had a significant financial interest in the Allhallows-on-Sea Estate Company who had acquired what is basically flat marshland. The intention was to convert the bleak, windswept area into a holiday resort to compete with Southend-on-Sea, situated in Essex, directly opposite the new station but on the far side of the Thames Estuary, as well as to develop a potential commuter hub.

Regular services started on the following Bank Holiday Monday, 16 May, local services shuttling to and from Gravesend Central with cheap day return services from Charing Cross being offered at 5/3d (approx. 25½p!), under-cutting tickets to any other resort. Commuters were enticed by two daily 'express' trains to Charing Cross, departing at 7.36am and 8.28am and returning early evening or around midday on Saturdays. These however were dropped by September, hopefully to return later when sufficient new homes had been built. Initially the popularity of Allhallows steadily increased with 9,500 passengers recorded as using the line on Bank Holiday Sunday, 5th August, 1934. Admittedly the beach, so to speak, was only 400 yards from the booking office with a tea room adding to the refreshment options available at the 'British Pilot', a Charrington's public house built in 1933. This sadly succumbed as recently as November 2021 and is much lamented by the author. In truth, traffic never really justified these facilities nor the significant goods yard and turntable. Only one daily pick-up freight was timetabled and it is

On the same day at Allhallows-on-Sea, Wainwright SECR Class H 0-4-4T No 31324 is the centre of attention for enthusiasts who have turned out to witness the last rites of the branch line. Introduced in 1904, those locos modified for push-pull operation in 1949 became the mainstay motive power for this and the Grain branch to the very end. The distinctive pillar water tank survived closure and demolition and now enjoys listed status within a mobile home complex. *A E Bennett 5886A / Transport Treasury*

reliably reported that the substantial goods shed never had any railway usage. For various reasons much of this infrastructure survived until demolition in 1975. The pillar water tank attained listed status and is now secure, together with a short section of platform, within a complex of mobile homes known as Kingsmead Park. In reality the anticipated development of the community never materialised with the area being noted as 'non-descript with few facilities' whilst the expected population 'explosion' between 1930 and 1960 went from 330 to just 580. However, until the outbreak of World War II, 12 return services operated daily to and from Gravesend increasing to 14 Down and 11 Up on Sundays. The Southern Railway at this point were considering electrification of services on this branch as well as the heavily-freighted Hundred of Hoo line but, against the increasing private ownership of motor cars, the future was doomed.

Rationalisation began with singling and push-pull operation by Wainwright's dependable Class H 0-4-4Ts which were the mainstay of the service until closure. To be fair British Railways did continue to promote the line and even tried out an ill-fated ACV 'lightweight' 125hp diesel railcar in 1953. Unfortunately, the noise and lack of comfort tended to turn passengers away, already aware that for a little extra money they could enjoy the delights of other resorts with much more to offer, such as Brighton and Margate. So with virtually no commuters using the line during the winter months, the writing was on the wall. Closure of course brought about hardship issues with the new bus route meeting general disapproval. The erstwhile Minister of Transport, the infamous Ernest Marples, declined to meet the local Conservative MP, Peter Kirk, but it is unlikely he would have given any creed to concerns over fog and ice affecting the very poor road network, never mind the additional cost and time of bus journeys. Ultimately, economy over-rode the question of public amenity.

So on the appointed day, the final 7-coach 8.38pm passenger train departed behind a grimy ex-SEC Class C 0-6-0 No. 31689, following clearance by Mr Solly. He was probably reflecting on the good times when the 'Flyer' would arrive with as many as 1,200 passengers from London every Sunday for a day trip to the seaside. He recognised the faces of his regulars but sometimes had to allow crowds of passengers off

Top: Having joined the main Hundred of Hoo Railway ay Stoke Junction, H Class No 31550 departs from Sharnal Street and heads towards Hoo Junction to join the North Kent line towards Gravesend Central. The single track through the white gate leads into the exchange sidings with the Chattenden Naval Tramway. This line also had exchange sidings at Lodge Hill with the 2'-6" Chattenden & Upnor Railway serving an extensive munitions depot. This is still referred to locally as the 'Chat-Up Line' and 'borrowed' in 1977 for the TV series 'The Upchat Line', starring John Alderton. ***A E Bennett 5888 / Transport Treasury***

Bottom: Alternative motive power awaiting departure from Allhallows with the branch service to Gravesend is provided by Tonbridge-based Bulleid 'Austerity' Class 5F 0-6-0 No 33031. Stabled alongside is an unidentified Birmingham RC&W Type 3 DE Bo-Bo (later Class 33) awaiting its next duty, probably the daily pick-up freight. ***HW 120 / Transport Treasury***

Top: Class H 0-4-4T No 31518 is awaiting right-away at Stoke Junction Halt to propel its 2-coach set around the curve to the left with a branch service from Gravesend across the wilderness towards Allhallows. *L N 2141 / Transport Treasury*

Bottom: Hither Green's Birmingham RC&W 'Crompton' Type 3 DE Bo-Bo No D6516 awaits departure from Sharnal Street with a mixed train for Grain. Behind the photographer is a notorious double-arched road overbridge, one for the Hundred of Hoo Railway and the other for the Kingsnorth Light Railway which, until 1940, ran through the exchange sidings to serve the Royal Navy airship factory. *L N 2142 / Transport Treasury*

in batches to avoid being overwhelmed. By the time of closure though he would hardly be over-stretched with around 30 passengers a day recorded, mainly going to work in Gravesend, school children or shopping trips. This is borne out by Rob Poole, a local resident who recalls the final days of the branch on a family outing from Gravesend to Allhallows in 1960. 'The ex-LSWR carriage we travelled in was far from clean. Accompanying us was our small, white terrier 'Bruce' who, after insisting on going under the seats, emerged in a greyish-black hue. Proceeding then to coat us with coal dust, all the family spent the rest of the day a somewhat lighter shade of grey, such were the delights of branch line travel in BR steam days.' He continued to consider the day's highlight to be an opportunity to travel on the short-lived miniature railway that then ran from alongside the 'British Pilot' to the resort's beach. No record can be found of this and any observations would be gratefully received.

Mr Solly, like others who worked on the branch line, had to consider offers of other jobs on the railway. This entailed moving from their village homes whilst a few were retained to handle the remaining goods traffic. Yet, from his point of view he only had 10 months left to serve and working at another station 'hardly seemed worthwhile'.

We can only wonder what Mr Solly would make of a new station currently being considered on the surviving freight-only Hundred of Hoo branch. Sharnal Street (or Hoo Parkway) may one day serve the ever-expanding local population, with electric trains to London running along the same tracks as his beloved 'Flyer'.

Many thanks to the Gravesend Railway Enthusiasts Society, particularly Ray Puddy and Rob Poole, for their co-operation in the preparation of this article.

Above: A unsuccessful attempt by BR to economise services finds ACV 'lightweight' 125hp diesel railcar waiting in vain for custom at the terminus on 24 October 1953. Observed in 3-car formation, this unit could operate singly or in 2-car mode but was deemed to be uncomfortable and noisy by passengers, used to padded seats and soot. The under-used goods yard appears to be hosting a rake of stored vans whilst the large building to the left is the 'British Pilot' public house, the hub of village life until closure in 2021. *R C Riley 4850 / Transport Treasury*

Opposite top: Unidentified Class H 0-4-4T approaches Stoke Junction from Grain with the 1¾ mile-long Allhallows branch curving away to the left. Behind the photographer is the first level crossing on the narrow and hazardous A228 to the Isle of Grain with a second crossing, by the entrance to the oil refinery, at Grain Crossing Halt which today still retains its manually operated gates. *A E Bennett 5870 / Transport Treasury*

Opposite bottom: At the other end of the branch we find Class H No 31324 shunting wagons in W. Weddel & Co.'s coal yard at Gravesend Central, having deposited its auto-set in the siding adjacent to the Down platform from which it will later propel its branch service to either Allhallows or Grain. Curiously the Ian Allan Combined Volume of the time does not indicate that this locomotive was push-pull fitted. *A E Bennett 5880 / Transport Treasury*

37
W. WEDDEL &

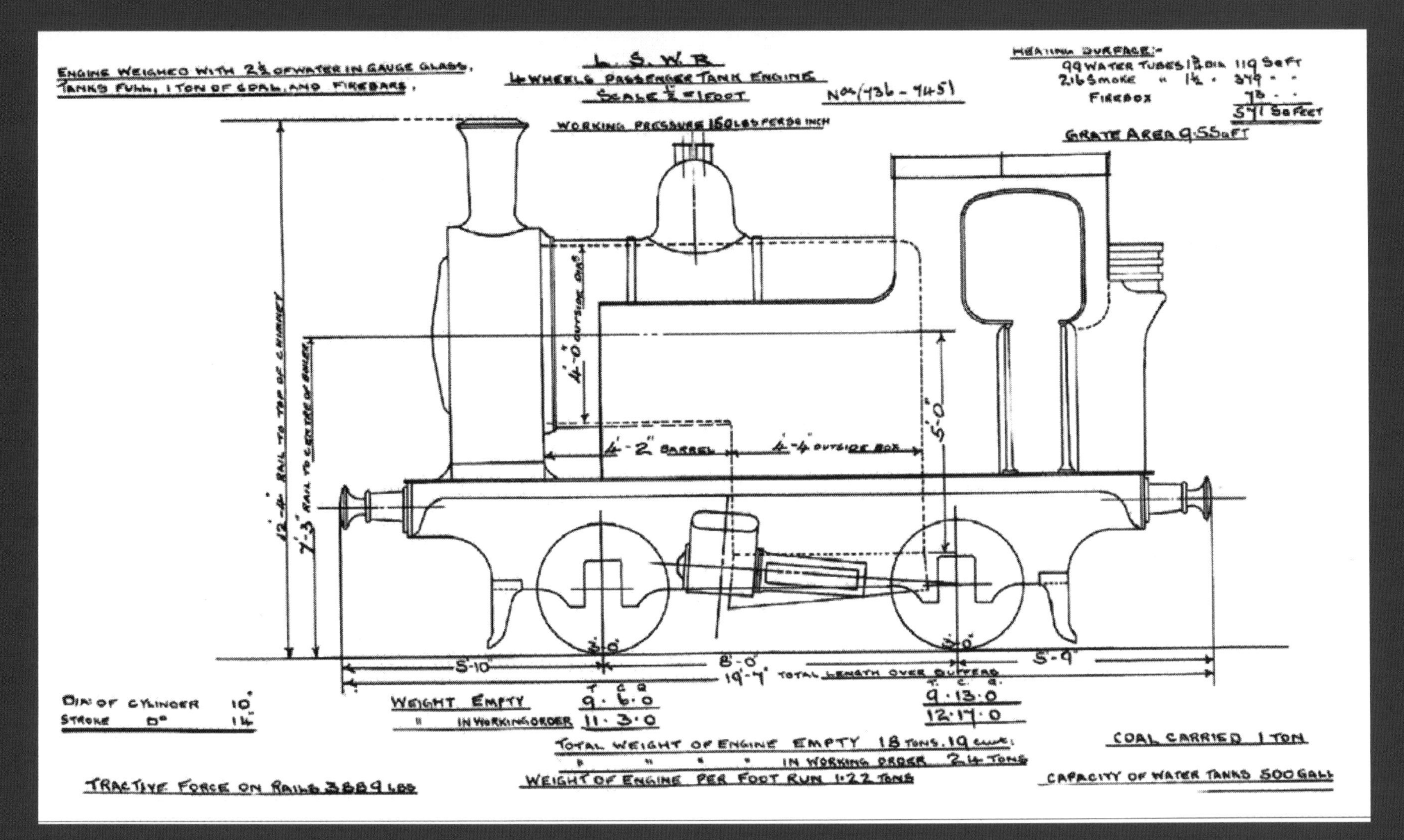

C14 class 0-4-0T

Mr Drummond's C14 and S14 Motor Tanks

In Issue No 1 of Southern Times we started our regular feature on locomotive types with a look at the various types of LSWR steam railmotor.

This time we turn our attention to their (theoretical) replacements, a series of small tank engines which in reality displayed all the disadvantages of the steam railmotors and few new advantages.

To start we should hark back briefly to the steam railmotor concept; a combined engine and carriage unit cheaper to build and operate than a separate locomotive and coach and one which could therefore reduce operating costs on certain branch lines. Fine in theory but this combination came with three disadvantages. Firstly in the event of the locomotive or carriage portion experiencing a failure then both were out of use. Secondly at times of high demand the railmotors were simply incapable of hauling an additional trailer coach and recourse had to be made to a conventional locomotive and rolling stock and in consequence negating any financial saving that might otherwise have been achieved. Thirdly was the need for servicing of the combined locomotive and coach unit at a steam shed where ash, cinders, dust and dirt were an inescapable aspect of the steam locomotive and which in turn would find its way into the passenger compartment of the combination.

Drummond's objective with the new design was to negate both the failure rate of having both loco and carriage out of commission at the same time whilst also avoiding the need for the carriage portion to venture into the steam depots and in these areas he succeeded. Where however he failed was simply that the power unit he designed was again woefully inadequate at hauling anything other than one coach.

The boiler for a C14 under construction at Nine Elms works. The 27 square-headed stays held the firebox cross tubes in place but would later be covered with the side tanks so making access difficult - see also subsequent captions.

Above: First of class again at Nine Elms and now in a more advanced state of completion. The motion is yet to be installed and also the cab roof, coal rails, dome cover, some pipework, couplings and the splasher over the front wheel. Even without the benefit of hindsight is must have been questionable as to the ability of a what was in effect a single wheel locomotive to haul two coach trains.

Opposite top: No 736 complete in works/photographic grey. A-top the cab the organ pipe whistle is visible along with the pulleys for motor-train operation. Sanding is obviously only provided to the driven axle. Before final completion Drummond was undecided on the best position for ownership and number decals and consideration had been given to these on the bunker side.

Opposite bottom: From a slightly more oblique angle we see the front end and as was fashionable at the time, the smokebox wing plates. Access to the firebox water tubes was both time consuming and difficult requiring the removal of the tank sides. The absence of any steps on the front framing would have meant access to clean the smokebox was not ideal. It appears couplings have yet to be added.

Notwithstanding what was perhaps obvious, he went ahead with a design for a diminutive 2-2-0T tank engine from Nine Elms, ten engines ordered in March 1906 at a cost of £910 each. Numbered 736 to 745, all but the last two were in service before the end of 1906 with the final pair following early in 1907. The cylinders, wheels and motion were identical to the H13 series of railmotors, whilst in an effort to increase hauling ability the type were given larger boilers having a total of 571 sq ft of evaporation surface including 119 sq ft of firebox water tubes. One ton of coal could be carried together with 500 gallons of water. As will be seen from the accompanying images, the boiler was high pitched whilst allied to the very short wheelbase the whole gave the impression of being squat and almost square in shape when comparing length with height. Pulleys were also added to the roof to facilitate cable attachments for the LSWR motor train system.

Trials with all ten engines took place around Eastleigh and Bournemouth after which they were designated to various local services, some of these previously having been operated by steam railmotors with some duties being shared.

These included Exeter - Topsham and occasionally Exeter - Honiton (Nos 736 & 740), Bournemouth West – Christchurch / New Milton and Bournemouth Central – Hurn – Ringwood (Nos 737 &744), Whitchurch –

L S W R
736
L S W R
736

L S W R
741
WHITECROSS

SOUTHERN
3741

Opposite top: No 741 in service livery. Likely taken when the engine was stored, changes include the provision in a tool box in lieu of the wing plate extension, removal of the roof mounted pulleys. The meaning of chalked inscription is not known.

Opposite bottom: The same engine but now as SR No 3741 at Eastleigh with another diminutive design No 0458 Ironside. (None of the C14 type were ever given official names.) By this time other changes have also taken place, the most obvious being the engine is now an 0-4-0T with the position of the cylinders moved to the front of framing. Handrails and footsteps have also been added at the front whilst there is also a vertical cabinet against the rear of the cab – crew lockers perhaps?

Above: Recorded at Eastleigh with the typical vista of the houses of Campbell Road behind – notice the 'E' prefix mid way on the tank side. A conventional whistle is now fitted and the clack valve is also visible – previously hidden by the framing around the smokebox. Finally to mention is the coupling has been replaced with a simple 3-link affair.

Hurstbourne – Fullerton, sometimes extended almost as a circular route to Andover Junction and then back to Whitchurch; note this was a service using ordinary and not motor train stock (739 & 742), Plymouth Friary – St Budeaux and occasionally to Turnchapel (Nos 741 & 743), Cosham – Havant again with conventional stock, (Nos 738 & 743). Note: Bradley gives the above details in his revised (Wild Swan) book on Drummond locos compared with the routes / services referred to in the original RCTS volume. The earlier addition referred to Southampton Town – Winchester (No 741), Portland Branch (No 744), Botley – Bishops Waltham (No 742), Brockenhurst – Lymington (No 743), and Plymouth Friary – St Budeaux (No 745). Three omissions are the Basingstoke – Alton line, Bishops Waltham, and likewise Ringwood to Christchurch all of which had previously seen railmotor use.

In service the fears expressed at the start were experienced almost immediately. Notwithstanding the increased boiler capacity they again proved incapable of hauling other than a single coach whilst slipping was a major issue at speed even where there was no gradient present and with consequential motion damage. Reliability was also poor, leakage from the firebox water tubes a frequent reason for the engine being classified a failure whilst access to these involved the removal of the side tanks, a difficult and time consuming operation. Four of the class, Nos 737, 739, 742, and 745 also received new cylinders within their first year; Bradley does not give the reason for this but slipping and water being carried over might seem the most likely cause.

Notwithstanding their obvious shortfalls, the class continued to perform on the task for which they been

Sister engine No 3744 again at Eastleigh. The pulley etc in front of the cab are nothing to do with the engine and relate to activity in the background. Note too the space between the coal rails has been filled in.

built that is until 1911 after which larger engines started to be used and the C14's were instead relegated to shunting at places such as Winchester, Southampton, and Dorchester. Six of the class though continued to work, Nos 737 & 740 on the Cosham – Havant shuttle and 738 & 740 between Bournemouth Central and New Milton. The fifth engine, No 742 continued on the Whitchurch – Fullerton service alternating with a member of the O2 class. Nos 736 & 741 also now found employment shunting Southampton Town Quay, the others stored at the rear of Eastleigh and apparently only steamed very occasionally - we are not told why such steaming took place, possibly simply to ensure they could be used if the situation demanded.

One of the Town Quay shunters also had a limited passenger turn on a Saturday working a single coach from the Town to the West station ready to be attached to a Bournemouth – Waterloo service.

The following year, 1912, the position had altered, Nos 743 and 745 were assisting the engineers with building work extensions at Eastleigh Works, No 740 coal stage pilot at Bournemouth and Nos 736-739, and 741-742 stored in the old loco works at Nine Elms.

With the departure of Drummond and his replacement by Robert Urie, the decision was made to rebuild five of the class, Nos 739, 741, 743, 744 and 745 as 0-4-0T engines, the others to be withdrawn when they reached the stage of needing heavy repairs.

Work started with No 745 which emerged from Eastleigh on 26 April 1913 with new 14" x 14" outside cylinders in line with the smokebox and a firebox without water tubes. Other necessary changes were new front wheels – as a crankpin was now needed – and of course coupling rods. A second engine No 743 was similarly modified in June 1913. No others were completed before the start of WW1 which gives the impression then that the rebuilding was hardly a priority.

Later there were other movements and also some hiring out of engines to third parties. No 743 transferred to Guildford in late 1914 and No 745 hired to Dixon & Cardus Ltd at Northam from February to June 1915. Around the same time No 738 was noted working at Tongham Gasworks, and No 744 to a contractor near Basingstoke – it was serviced at Basingstoke on Sundays.

WW1 however did provide a brief restoration of work for the class for in March 1915 Nos 741 and 744 were engaged with a trailer car transporting workman between Bournemouth West and Holton Heath where a munitions factory had been established at the latter location. Considering how the original 2-2-0T design had fared slipping on a level gradient it would be interesting to speculate how they might have fared on the return working tackling the formidable Parkstone Bank; although we have no information from the time it is probably to surmise by saying 'not well'. Two others, Nos 736 and 739 were loaned to the Docks Department at Southampton whilst No 737 could be seen shunting at Winchester.

The next stage commenced on 12 November 1916 when the War Office approached the LSWR to ascertain if any light locomotives might be available and in return were offered seven members of the class, this stage is best described in the below table:

736	Sold Ministry of Munitions March 1917 for £1,050	Worked at Royd's Green, Leeds until November 1917. Then transferred to Shoeburyness (Royal Engineers Garrison Railway).
737	Sold Admiralty December 1917 for £1,100	Worked at Mine Department, Grangemouth.
738	Sold Ministry of Munitions March 1917 for £1,050	Worked at Dunball near Bridgewater.
739*	Sold Bute Works Supply Company February 1917 for £900	Worked at H M Ordnance Factory, Bramley
740	Sold War Department December 1916 for £900	Used by Inland Waterways & Docks, Richborough, transferred to Shoeburyness in October 1917
742	Sold Ministry of Munitions March 1917 for £1,050	Worked at Shoeburyness.
743	On loan to Portsmouth Dockyard from 23 March 1917, then sold to the Admiralty November 1917 for £1,100	Worked in Portsmouth Dockyard.

Ownership of two others of the class Nos, 741 and 744, was retained by the LSWR but they were hired to the Admiralty.

No 741* worked at Portsmouth Dockyard from 17 July 1917 until 26 November 1919 whilst No 744 was at the Royal Navy Depot at Bedenham (Gosport) from 11 January 1917 to 23 February 1920.

* At some stage Nos Nos 739 and 741 had been modified with 11" x 14" cylinders and with the removal of the firebox cross water tubes.

Motion detail from No 30589 at Eastleigh on 15 August 1956. Former Eastleigh apprentice Mark Abbott recounts this was the only class of steam engine from which he could remove and replace the motion single handed. This is the former LSWR No 744, SR No 0744/3744.

Top: In lined out BR black livery, No 30588 (the former LSWR No 741, SR No 0741/3741), again at Eastleigh. *Neville Stead / Transport Treasury*

Bottom: Locomotive and crew posed on the Town Quay at Southampton, which set of rails ran from the old Eastern docks to the newer Western docks. The firm of Edwin Jones a department store whose retail premises were near the Bargate, 26 June 1957. *R C Riley 10852A / Transport Treasury*

Top: A rear view of the third of the surviving trio, No 77S at Redbridge on 26 June 1957. Shunting duty could clearly be perilous for anyone required to hang on to the side of the engine…. . *R C Riley 10863A / Transport Treasury*

Bottom: No 77S marked here as allocated to the Engineer's Department and correctly stabled in mid gear at Eastleigh. Manipulating fireirons – which appear to have been kept o top of the bunker – cannot have been easy. The height of the safety valves on the dome appears to vary with this engine compared with her sisters. Bradley records only one mileage for a member of the class, No 77S purported to have run 321,643 miles but this must be open to question *Neville Stead / Transport Treasury*

3210
30588

Opposite top: No 30588 amongst the opposition at the Southampton Terminus servicing point. (The two WR engines would have worked trains from their region to the town.) The 0-4-0T may well have been present to use the turntable. Serving for the engines working at the Town Quay was likely to have been at Eastleigh.

Opposite bottom: In its home environment, No 30589 shunting the sidings near the Royal Pier. AS may be seen the public could also walk nearby with no separation between road and railway. In the background are the town walls. Latterely this engine was running with a B4 chimney and it will also be noted without tool boxes on the front framing. (Southampton did not become a City until a few years later and in consequence of this article is thus correctly referred to as being a Town.)

Above: No 30589 at Bishops Waltham on the occasion of its 1952 outing, This is believed to have been the only passenger working by a member of the class for almost 30 years.

Of the seven engines identified in the table, all excepting No 739 later came under the control of the Government Disposals Board and post WW1 were offered for sale being described as 'A new type of light passenger or shunting locomotive', the description was accompanied by an illustration of No 736 in LSWR livery. – possibly that seen on page 20 bottom.

Three, Nos 736, 740 and 742 were purchased for scrap by James Brown Ltd of Sittingbourne and broken up, whilst Nos 738 and 743 were shipped to India from Glasgow in February 1923. The final one of the six, No 737 was noted at Woolwich Arsenal being used to supply steam to test the boilers of N class 2-6-0 tender engines then being built there for the SECR. Like its larger compatriots of the N class it too found its way to Ashford where it was officially noted as 'small stationary boiler' No WA740. It was scrapped at Ashford in May 1924.

This left three engines out of the original ten. No 745 rebuilt as an 0-4-0T had been regularly employed shunting at Southampton Town Quay after 1914 but in August 1917 was working on the Lee-on-the-Solent Railway and in March 1919 at Winchester. No 741 was transferred to Strawberry Hill for engineers work tipping soil and ballast in connection with the building of the new yard at Feltham. It was joined by Nos 744 after the latter had completed its turn at Bedenham both engines still in their original 2-2-0T form. In this guide they were of limited use and consequently were rebuilt as 0-4-0T's identical to No 745 in March 1922 and October 1923 respectively.

Meanwhile their numbering had also altered, all three now being on the duplicate list having a '0' added in from of the original number. No 0745 was further renumbered 77S in October 1927 and transferred to the Service Department for work at Redbridge Sleeper and Permanent Way Depot.

Initially under SR ownership all three were bestowed with lined black livery but this gave way to plain black as subsequent economy measures took place.

Nos 0741 and 0744 were renumbered as 3741 and 3744 in 1931 retaining this identification until BR days when they became Nos 30588 (December 1950) and 30589 (August 1948). Under BR all three survivors reverted to lined black livery whilst before this handrails and footsteps were provided at the smokebox end of the engines for the use of shunters, tool boxes were affixed to the running plate above each cylinder. Bradley comments that new copper fireboxes had been supplied in 1947-51 (without further detail), whilst each engine also retained its own boiler and consequently if called for repair as no spares were available.

Almost 40 odd years since of the type had been used on a passenger duty, No 30589 was especially selected for an RCTS two-coach special train from Eastleigh to Botley / Bishops Waltham on 14 June 1952. The engine was especially cleaned and apparently reached a speed of 35 mph, a not inconsiderable amount for an engine with driving wheels just 3' in diameter.

Five years later came the end for the two engines in capital stock, Nos 30588 and 30589 laid aside in December and June 1957 respectively. Their replacement was a small 204hp diesel shunter but their were manning issues over its use and 77S was instead requisitioned from Redbridge and took over the same Town Quay duties until it too was withdrawn in April 1959.

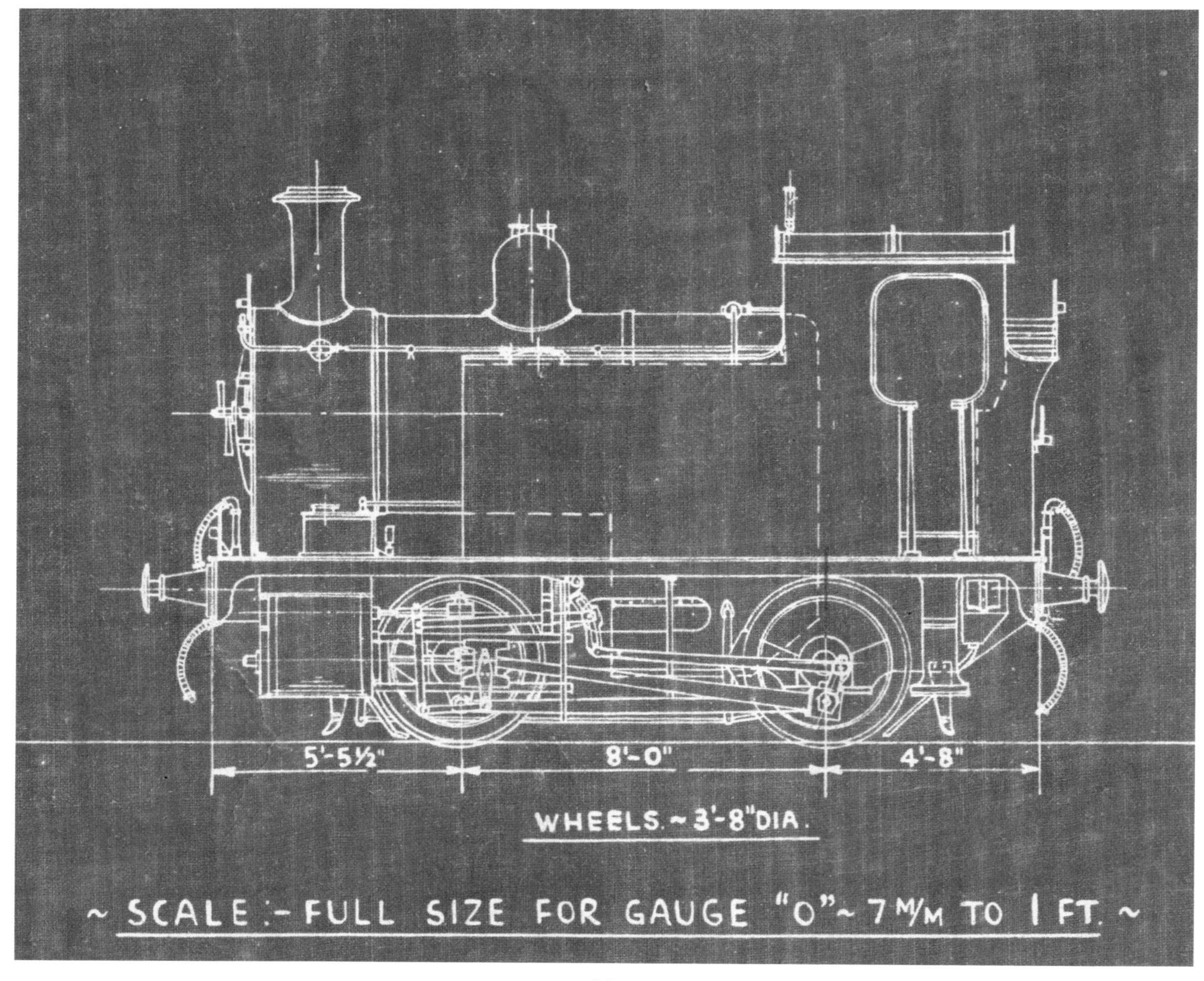

The S14 0-4-0T

Similar in appearance to but slightly enlarged, were a further five 0-4-0T engines ordered in June 1908 and intended to be numbered 746 to 750 of class S14. These had coupled wheels 3' 8" in diameter, an increased heating surface and other differences thereby increasing the weight to 28 1½T. Evidently the final design had been the subject of some debate between Drummond and Urie, the latter preferring an 0-4-2T design with the inherent advantage of smoother running and increased coal and water capacity. Drummond's view however was that adhesive weight would be lost with a set of carrying wheels whilst he was adamant that the new design would cope with two and three coach trains.

In the event just two engines were built, with the balance of the order cancelled in August 1910 at which point assembly of the first two was well advanced and completed in September 1910. Boilers for the other three had already been finished and suitably modified instead found other use, two providing power for the electric generators at Waterloo and the third for a well pump at Micheldever.

The two completed engines appeared with revised numbers, 101 and 147 and whilst a definite improvement on the smaller C14. Drummond's prediction of the design being able to haul three-coach trains again proved overtly optimistic. The short wheelbase also meant the engines rode badly both when accelerating and at speed and the engines quickly became unpopular with crews and passengers who were subjected to a similar uncomfortable ride.

Work wise No 101 was dispatched to work trains on the Lymington branch, No 147 on the Ringwood – Bournemouth line via Hurn. No 101 joined No 147 on the same duties after a short time. Trials on the Portland and Seaton branches were unsatisfactory. By November 1916, No 101 was reported as shed pilot at Guildford, No 147 presumably continuing its work on the Hurn branch.

The pair's time on the LSWR was destined to be brief, for in March 1917 both were sold for £1,250 each to the Ministry of Munitions. No 147 was sent in steam to a munitions depot at Quedgeley on 17 April and later to Trumpington Ordnance Depot, Cambridge. After some necessary repairs at Eastleigh, No 101 departed for Stratton Filling Station near Swindon on 19 May.

Eastleigh provided some (unspecified) spare parts for

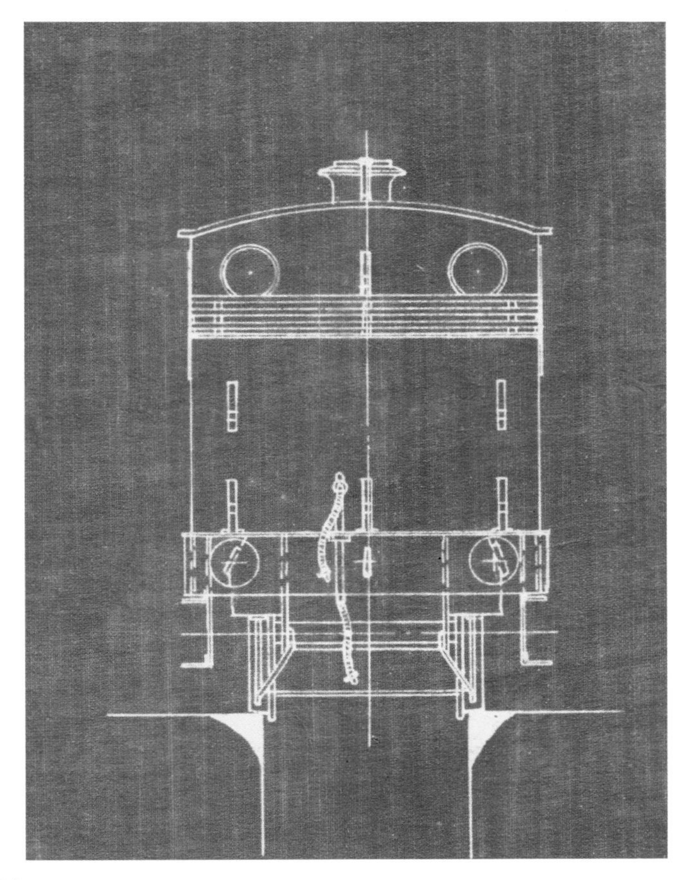

No 147 and again in June 1918. In 1920 it was offered for sale and purchased by Joseph Pugsley & Sons who despatched it to the Avonside Works at Bristol for conversion to oil firing. It was next heard of at Swansea Docks on 20 July 1927 for shipment overseas after which no further information is available.

We know slightly more about No 101 as this was sold to the Branstone Blue Lias, Lime & Cement Company Ltd in October 1921 and who subsequently obtained some spares from Eastleigh In March 1925. It was next reported at the yard of G Cowan, Stanningley, Leeds on 7 August 1939 after which it was repaired at Derby Works still inscribed G Cohen & Sons having arrived there by road with chimney and cab removed. Following repair it travelled by rail – not known if under its own steam – to Grangemouth where D L Bradley noted it as shunting in April 1944. Seven years later in 1951 it was noted as a stationary boiler at a Royal Ordnance Factory at Chorley. Apart from the removal of the vacuum brake pipes it appeared complete but was awaiting scrap.

So ended the saga of the Drummond 2-2-0T and 0-4-0T types. Underpowered and basically unsuitable for their original design purpose several survived to provide service far away from their original haunts. None would make it to preservation.

Above: One of two examples of the slightly larger S14 0-4-0T locos built. This is No 101 when new. The larger size coupled wheels precluded the provision of splashes, otherwise there is marked similarity between the two types.

Opposite: The second of the class recorded alonside the original station at Christchurch with a service to Ringwood. It is likely in their brief spell of existence the design adopted the same nickname as that of the C14, namely 'humpy-dumpties'.

Next time - the J class 0-6-4T engines of the SECR

L S W R
147

Recollections at Exeter Central
Saturday 20 July 1963
John Bradbeer

On the rear cover of ST7 we featured an image of No 34058 preparing to take over the London bound ACE at Exeter Central. This resulted in the following letter from John Bradbeer along with the accompanying table; we just hope the associated may bring back a few memories from John and others.

'I guess that you have readers far more familiar with Exeter Central operations than I but it struck me that the picture was probably of one of the Saturday portions of the ACE. The usual chain of events saw the North Devon portion arrive first. Passengers would disembark and the train then draw forward clear of the scissors crossover. The North Devon portion's engine would then run light to Exmouth Junction. The station pilot would attach the catering vehicles and any others to the rear of the North Devon portion before the Plymouth and North Cornwall portion arrived. Its locomotive would use the scissors cross-over to clear its train and similarly run light to Exmouth Junction. The train engine for the Waterloo run would have arrived light engine from Exmouth Junction and back the North Devon portion and the catering vehicles on to the Plymouth portion. This picture seems to show No 34058 having arrived light engine (the 'tail lamp is on the buffer beam) and is collecting some Bulleid coaches from the middle road and probably will use the scissors cross-over to run into the platform line ahead of the first portion to have arrived. Now, on Saturdays, the ACE from Ilfracombe ran with its own catering from Ilfracombe, and did not attach anything at Exeter Central. The Torrington portion did not have catering and I wonder if 34058 about to attach a few more carriages and the catering vehicles to the Torrington portion, which may either be hidden behind the train in the middle road or yet to arrive? I think that there was no catering either on the Padstow and Bude portion, so it could be this rather than the Torrington portion to be taken over by 34058. I grew up in Barnstaple and my father's brother-in-law was a signalman at Exeter Central A box, but on our visits to Exeter we caught an earlier train than the ACE, so I would not have seen multiple ACE's that often.'

Arrivals		
Time	**Locomotive**	**Working**
10-11 up	34106, 31914(b)	08-10 Ilfracombe-Waterloo
10-21 down	41206	09-54 Exmouth-Yeoford
10-25 up	82002 (p), 34002	08-25 Plymouth-Waterloo
10-39 term	80039	10-05 Honiton-Exeter Central
10-48 up	34072, 31914 (b)	08-35 Ilfracombe & 09-10 Torrington-Waterloo
10-53 term	35006	08-00 Salisbury-Exeter Central
11-06 down	34082	07-28 Waterloo- Padstow & Ilfracombe
11-10 term	80040	10-44 Exmouth-Exeter Central
11-23 down	35018	Surbiton-Okehampton car carrier
11-26 up	34065	08-10 Wadebridge & 09-00 Bude - Waterloo
11-33 up	34070, 31914(b)	10-00 Mortehoe- Waterloo
11-45 term	DMU	11-13 Exmouth-Exeter Central
11-50 up	34076, 31911(b)	08-30 Padstow & 09-30 Bude – Waterloo
12-06 up	31835	10-02 Plymouth-Portsmouth & Southsea
12-11 term	41306	11-45 Exmouth-Exeter Central
12-20 down	35019	08-35 Waterloo-Ilfracombe & Plymouth
12-21 up	34078	10-30 Ilfracombe-Waterloo (ACE)
12-35 term	41321	12-08 Sidmouth Jtc-Exeter Central
12-39 up	34079, 31914(b)	10-48 Torrington-Waterloo (ACE)
12-44 term	80036	12-15 Exmouth-Exeter Central
12-48 down	34067	09-03 Portsmouth & Southsea – Plymouth
12-52 up	34060	11-10 Plymouth-Brighton
13-17 term	41206	12-47 Exmouth-Exeter Central
13-38 down	34024	10-15 Waterloo-Ilfracombe & Torrington
13-42 term	DMU	13-15 Exmouth-Exeter Central
13-45 up	34081	12-00 Ilfracombe-Waterloo
13-55 down	34049	10-35 Waterloo-Padstow (ACE)
14-04 up	31856 (p) 34011	11-00 Padstow-Waterloo (ACE)
14-11 term	41306	13-45 Exmouth-Exeter Central
14-19 term	41309	13-45 Honiton-Exeter Central
14-20 up	D6342	11-48 Plymouth-Waterloo

Busy times at Exeter Central. On the far left we have No 34086 *219 Squadron* displaying on the tender the code for light engines between Exmouth Junction and Exeter Central. Meanwhile No 35008 *Orient Line* runs light back to the depot having brought a train in from Waterloo – the note the light grey hue in the chimney indicating good combustion. On the right a rebuilt West Country, No 34096 *Trevone* displays the headcode for a Plymouth service.

14-28 down	35022	11-00 Waterloo-Ilfracombe & Torrington (ACE)
14-36 down	35020	11-15 Waterloo-Plymouth, Padstow & Bude (ACE)
14-41 term	80036	14-15 Exmouth-Exeter Central
14-50 up	31874	12-40 Torrington-Waterloo
15-06 term	30842	14-08 Axminster-Exeter Central
15-15 term	DMU	14-49 Exmouth-Exeter Central
15-30 down	34034	12-05 Waterloo-Ilfracombe
15-47 down	34086	12-15 Portsmouth & Southsea- Ilfracombe & Torrington
16-01 term	30841	15-21 Seaton Junc-Exeter Central
16-04 term	80035	15-36 Exmouth- Exeter Central

b - banker

p - pilot

John had arrived at Exeter on the 08-54 departure from Barnstaple Junction

Opposite bottom; Shunt move. No 34005 *Barnstaple* is leaving the down platform and is crossing on to the up main with a pair of Maunsell vehicles. With the West of England headcode displayed it is tempting to suggest this might be an eastbound stopping service but in reality perhaps unlikely although (from another image taken shortly after) we know No 34005 continued east beyond the station.

This page, top: No 34086 again, now with approximately seven coaches, all excepting the first, Bulleid stock. Likely this is indeed a Salisbury or Waterloo service.

This page, middle: In the opposite direction, No 34026 *Yes Tor* covers the short distance from the tunnel to Exeter Central with a down working.

This page, bottom: No 35009 *Shaw Savill* Waterloo bound. With its load of 12 or so coaches, this would be the combined portions of services from the 'Withered Arm'.

SOUTHERN RAILWAY

Signal Instruction
No. 10, 1940.

Instructions to all concerned as to

INTRODUCTION OF COLOUR LIGHT SIGNALS BETWEEN CLAPHAM JUNCTION "E" AND POINT PLEASANT JUNCTION SIGNAL BOXES

(In place of existing semaphore running signals)

AND

ABOLITION OF EXISTING WANDSWORTH TOWN SIGNAL BOX,

ON SUNDAY, 26th MAY, 1940.

Rules 77, 78, 79 and 80 to be observed. Drivers to keep a good look-out for hand signals.

Commencing at 1.0 a.m. on Sunday, 26th May, the existing semaphore running signals on the down and up lines between Clapham Junction "E" and Point Pleasant Junction signal boxes will be abolished and colour light signals installed in lieu thereof.

The existing Wandsworth Town signal box will be abolished.

The new signals and their location are shown on the diagram accompanying this Signal Instruction. The signals prefixed by the letters WBB will be worked from Clapham Junction "E" signal box and those prefixed WBD from Point Pleasant Junction signal box.

No alteration will be made in the application of the existing shunting signals at Clapham Junction "E" and Point Pleasant Junction.

A plate bearing the prefix letters and the number of the signal will be fixed to each colour light signal post.

The colour light running signals will show four or three aspects.

The lights of the four-aspect running signals will be arranged as shown on the diagram and not as appearing on page 4 of the General Appendix to the Working Time Tables.

The aspect of the colour light running signals will be the same by day as by night.

Colour light running signals will be fitted with small side lights repeating the aspects exhibited by the signals to assist Drivers of trains drawn close up to such signals.

Back lights will not be provided in any of the colour light running signals.

The height of the centre of the red light of the colour light running signals will vary between 9 and 17½ feet above rail level.

Track circuits have been installed throughout the area covered by the colour light signals, and all colour light running signals, except Point Pleasant Junction down home signals, will be controlled by the track circuits.

The colour light running signals, except Point Pleasant Junction down home signals, will be replaced to Danger after the engine has passed a distance varying from 15 to 200 yards beyond the signal.

New 3-aspect approach light signals will be provided beneath Point Pleasant Junction up home signals. When the semaphore arm is at Danger a red light will be exhibited in that signal and no indication will be shown in the approach light signal. When the semaphore arm is lowered, a green light will not be exhibited in that signal but a yellow, two yellow or green aspect will be shown in the approach light signal.

JUNCTION INDICATORS.

Junction Indicators will be provided at certain signals as shown on the diagram and will apply as indicated in Rule 35, clause (e).

TELEPHONES.

Telephones will be provided at or adjacent to certain signals as indicated on the diagram.

SIGNALLING DURING FOG OR FALLING SNOW.

Fogsignalmen will not be provided at any of the colour light signals referred to in this Signal Instruction.

H. E. O. WHEELER,
Superintendent of Operation.

Deepdene Hotel,
Dorking.

17th May, 1940.

(R. 62030.)

Waterlow & Sons Limited, London Wall, London.

Visits to Point Pleasant signal box in the early 1970s

Les Price

Some years back in the editor's predecessor to 'Southern Times'; I read Richard Simmons postscript to his article 'The Lines and Stations Dr Beeching did not close'. It took me back almost a full fifty years to the early 1970s. At the time I was a 'Bobby'. Not one of those formerly connected to the description of a signalman or 'Constable on the track'; but one of Robert Peel's men in the Metropolitan Police whose patch was the streets of Wandsworth in south-west London. Even as late as 1973 'Beat Policing' was still very much in vogue and one of the Wandsworth 'Beats' extended along the southern bank of the Thames all the way via Putney Bridge Road to Putney High Street itself.

At that time the Met. stabled some of its police horses at Wandsworth and if one walked down the cobbled slope from the stables and back yard of the Police Station to the road below and turned left you very quickly came to the 'Queen Adelaide' pub; one of a chain owned by 'Young's' Wandsworth Ales; an excellent pint! Every 'Copper' worth his salt would know exactly where the pubs were on his beat; they were often the centre of his problems! The writer can accredit the excellence of Young's Ales due to 'Night Duty' visits to check on the safety of the night watchman at the Brewery. Way back in 1973 it usually resulted in a visit to the 'Tap Room'. But that's another story!

Directly opposite the 'Queen Adelaide' was Point Pleasant Road, which ran through an industrial area down towards the river, not that the road ever arrived there. The name of this place was assumed to have been taken from a comment once ascribed to Queen Victoria. Early in her reign she visited this area on the Thames when it was still quite rural and described it saying 'this is a pleasant point".

A short distance along Point Pleasant Road, on the right hand side, just before a railway overbridge there was a gate. Being an inquisitive young Bobby wanting to explore every inch of his 'Beat' and needing for future reference to know where the path led; one day I went through it to investigate. It ended with climbing the steps up to a Signal Box.
I knocked on the door and the signalman who opened it was a small featured dapper, congenial Irishman by name of Pat Cotter. He invited me in and put the kettle on. Thus was my introduction to Point Pleasant Junction Signal Box.

It has often been recorded that a Driver or his Fireman, held at a signal, would go to the lineside telephone enquiring, 'What's up, Bobby?' This persisted right through to the dying days of steam. Hence the association was made.

The connection was forged over a mug of tea. In historic terms we had much in common. His roll, as a 'Bobby' was to police the rails i.e. the movement of trains; I was supposed to perform a similar function on the streets and in the community keeping the wheels of society quietly turning. He welcomed my interest in railways with open arms, in particular his job as a signalman. After a century and a half our individual roles had significantly changed but the kinship seemed to remain.

Even after this time lapse there was still an affinity, albeit indistinct. Here was precisely where the two roles came together so I did not feel out of place in that signal box. I felt a connection with one of my distant ancestors, a signalman on the Cheshire Lines Committee Railways at Northwich, in the late nineteenth century. Strange how family connections are carried forward.

I found out the East Putney cord ran across the previously mentioned bridge. It had been built in 1886 by the District Railway as a double track line from their river crossing to a junction with the Windsor Lines at Point Pleasant, which was intended to give them access to Waterloo. But it also allowed an alternative route to Wimbledon for the London & South Western Railway to complement their main line via Clapham Junction. Thus Point Pleasant Junction signal box was opened by the L&SWR in 1889.

It was built to their Type 3b design and fitted with a 46 lever Stevens & Sons Tappet frame (manufactured by The Railway Signal Company Limited). In 1940 the Southern Railway replaced the original box with a Type 14 'Air Raid Protection', (ARP) Signal Box.

Aesthetically the Signal Box was less than pleasing. Unlike many boxes I had known, which effortlessly blended into their surroundings, Point Pleasant was squat and austere, the very antithesis of attractive. It was similar in appearance to Crewe Bank Signal Box at Shrewsbury, which I remembered from my boyhood. I soon found out the reason for this design. Both were virtually identical and of a group of such boxes built under wartime conditions during the Second World War with greater protection against bomb damage.

Such new ARP Boxes were built throughout the country during the early years of the war. On the Southern Railway the ARP design was similar to a new design previously introduced in the second half of the 1930s as exemplified by the then new signal box at Templecombe opened on the 15 May 1938. However the wartime ARP version featured square rather than curved corners to the building and a more substantial roof with no overhang.

Point Pleasant was one of around thirty of such designs to be built between 1940 and 1949. It also had sliding windows as distinct to the traditional opening variety. In the case of Point Pleasant Junction a new brick structure was built around the original wooden box and then the upper external structure of the old box was demolished from inside but the original structure inside the new box remained.

With the onset of war the importance of the East Putney line clearly took on a new significance. It provided a by-pass in the event of air raid damage or congestion on the main line between Clapham Junction and Wimbledon. Southern Railway Signal Instruction No.10 of 1940 'Introduction of Colour Light Signals Between Clapham Junction 'E' and Point Pleasant Junction Signal Boxes' explains when the transformation to an 'ARP' Box may have taken place.

This work included the abolition of the existing 'Wandsworth Town Signal Box' and was commissioned to be carried out on the 26 May 1940. In the event this date coincided with the evacuation of the British Expeditionary Force from France which commenced on the same date. The Southern Railway then had greater priorities, organising trains at South Coast ports to convey injured and demoralised troops to hospitals and camps throughout the country. In consequence the

The interior of the signal box looking towards Putney, showing the Box Diagram and frame with Signalman Pat Cotter checking the progress of an 'Up' train on the Diagram. The East Putney Lines can readily be seen, 11 April 1973. ***All images by the Author***

Top: Under the watchful eye of Signalman Pat Cotter, another fairly comprehensive view of the frame this time looking east towards Clapham Junction.

Bottom: On Thursday 12 April 1973, Class 73 No E6003 descends from the 'Up' East Putney cord at the head of the 13.07 Morden milk empties heading for 'Kenny Sidings' at Clapham. The Express Dairies creamery at Morden received much of its milk overnight from Torrington. The empties would be returned by this train to Clapham to be assembled into the afternoon milk empties to Salisbury and the West Country which itself was routed via the 'Down' East Putney chord to Wimbledon. Some tanks bound for other destinations off this train may also have been added to 6V12 which left Clapham Junction at much the same time for Kensington Olympia. The Code G3 indicated that the train was routed via East Putney. No E6003 entered service from Eastleigh Works on 27 April 1962. As a result of the introduction of TOPS it was officially renumbered as 73003 on 1 January 1973 but was still sporting its original number at the time of this photograph. It was withdrawn from service in September 1996 and is now preserved in private ownership.

signalling work was not carried out until 16 June, two weeks later.

The East Putney cord from Wimbledon to Wandsworth Town was used by the Southern for empty stock movements and occasional service train diversions, as well as some early morning trains to and from Waterloo for train crew. Another use was to route milk trains and the return empties as a useful means of getting between Wimbledon and Clapham Jnc. West Yard where they were sorted.

It continued to control this traffic until closed on 16 September 1990 when its area of responsibility passed to Wimbledon Area Signalling Centre. It was the last surviving 'ARP' Box on the Southern Region. The 'Up' track from Wimbledon crossed over the tracks of the 'Windsor' lines via the bridge west of Point Pleasant Box before merging with those tracks at Point Pleasant junction. This link has long since been closed and the main deck of the viaduct has been removed.

After four years serving at Wandsworth and numerous visits to the Box the time came for me to move on. Clearly the call of the railways was too great and at the beginning of September 1976 I transferred to the British Transport Police at Crewe. It just so happened that my last tour of duty in London was at the Notting Hill Carnival on August Bank Holiday Monday, the first Notting Hill riot. The only protection from flying milk bottles and other missiles was a stoutly grasped dustbin lid. What could the BTP offer instead? Well that's another story!

Of course the standard daily fare passing to and fro at Point Pleasant was the ex-Southern Railway designed 4-Sub Units. In this view taken on 11 April 1973 a gleaming set No 4747, again carrying the G3 route indicator and obviously recently refurbished descends off the East Putney cord. It is clearly en-route from Wimbledon Depot to Waterloo to take up its scheduled duty. Number 4747 tells us it was one of the later sets built by British Railways at Eastleigh after 1949. During 1971 British Railways introduced the Total Operations Processing System (TOPS) to manage locomotives and rolling stock. Under this system 4-Sub units became designated Class 405. But this was divided into two sub-classes, 405/1 and 405/2. The former included those built during the Southern Railway era and the latter those built after nationalisation from 1949 to 1951. No 4747 was in this latter group numbered between 4601 and 4754; in fact it would appear to have been in the last batch. All had been withdrawn by 1983.

Opposite: Looking east, with Wandsworth Town station in the background, on Wednesday 11 April 1973, we have a signalman's view of what appears to be a Class 45 locomotive displaying Train Reporting Number 8D28. It is running west with the 08.00 Betteshanger Colliery - Brent coal train. The fascinating feature of this picture is that the train would have originated on one of Col Stephens' former railways, namely the East Kent Light Railway. Whilst on that line the services north of Tilmanstone Colliery ceased on 1 March 1951 but the southern stub between Tilmanstone and Sheperdswell remained open for coal traffic under the auspices of BR. Thus this train would have joined the Dover to Canterbury East main line at Sheperdswell and continued its journey along the North Kent main line towards Gravesend before probably striking off south-west via Swanley and Bickley Junction to follow a route across South London through Dulwich and Herne Hill to Brixton. From here it would likely have continued that intricate route to Clapham Junction before arriving here at Point Pleasant. Its journey onward would have been via Kew Bridge Junction on to the North and South-Western Junction line to Acton Wells Jnc. and Willesden before eventually arriving at Brent Junction. Did any other working ever follow such a convoluted route?

32106
32106

Stephen Townroe's Colour Archive In and out of Works

For this issue's selection, we turn our attention to the activities within Eastleigh Works (mainly) with a few excursions outside as well as one trip 'over the sea'. Stephen Townroe was a frequent visitor to Eastleigh works sometimes recording the engines, sometimes the heavy engineering, and make no mistake, railway workshops did involve heavy engineering.

Opposite: E2 0-6-0T No 32151 approaching the end of an overhaul at Eastleigh some time in the early 1950s and seen here in the position for valve setting. This vital task was performed when all the inside motion had been set up with adjustments made to the valve rods etc to ensure the engine responded to the position of reversing lever. No 32106 had been until under the auspices of L Billinton at Brighton in 1915 and spent its last years shunting at Southampton Docks. It was withdrawn after 47 years service in October 1962.

Above: Fairburn design LMS 2-6-4T No 42090 being lifted at Eastleigh in 1954. (Thanks to its nameplate, we can easily identify the Pacific alongside as No 34090.) The 2-6-4T was one of a batch of the class built at Brighton in 1951 for the Southern Region and so three years may be at the start of its first overhaul. Like so many other steam engines of the same era it was destined to have a unduly short life and was withdrawn in 1964 after just 13 years service.

This page: The former LBSCR No 54 Waddon, latterly DS680, seen here restored externally to LBSCR condition at Eastleigh prior to being shipped for static display at the Canadian Railway Museum in Quebec.

Opposite top: Quiet times within Eastleigh Works in September 1951 meaning this was probably recorded on a Sunday. Several engines are in various stages of assembly/ disassembly that nearest the camera a ‘700’ class 0-6-0T.

Opposite bottom: S15 4-6-0 No 30521 outside the works and fresh from overhaul in 1950. Although perhaps more akin to LMS rather than Southern colours, the lined black livery sat well these chunky 4-6-0s, the only item missing for the present being any form of insignia on the tender. The engine will likely soon be worked around to the shed where a trial steaming and test run will take place, light engine, probably to Micheldever. After that it might be ‘borrowed’ by Eastleigh for a day or so before being returned to its home shed. (Note; this image is canned right to the edges, meaning the extremities of the locomotive are missing from the original scan.)

30521

This page, top: Away from the completed engines, we now turn our attention to the activities with in the works, such as here with tyre being heated by a gas ring ready for it to be cooled and in the process shrunk on to the wheel – evidently a Bulleid wheel.

This page, centre: Bulleid crank axle wheel set from a Pacific complete with sprocket to drive the chain driven valve gear. 1951.

This page bottom: The wheel shop in 1954 with a pair of wheels from a Merchant Navy in the balancing machine, no guard of any sort appears to be provided.

Opposite top: The noisiest part of the works, the boiler shop, here a boiler from a 4-6-0 is in position for stay drilling, 1954.

Opposite bottom: Bulleid boiler and smokebox in the boiler shop, 1952/3.

Overleaf: Display of brass items ready for a visit by members of the Institute of Mechanical Engineers to Eastleigh in 1955. We would readily point out the crests at either end do not go with the *Seaton* nameplate!

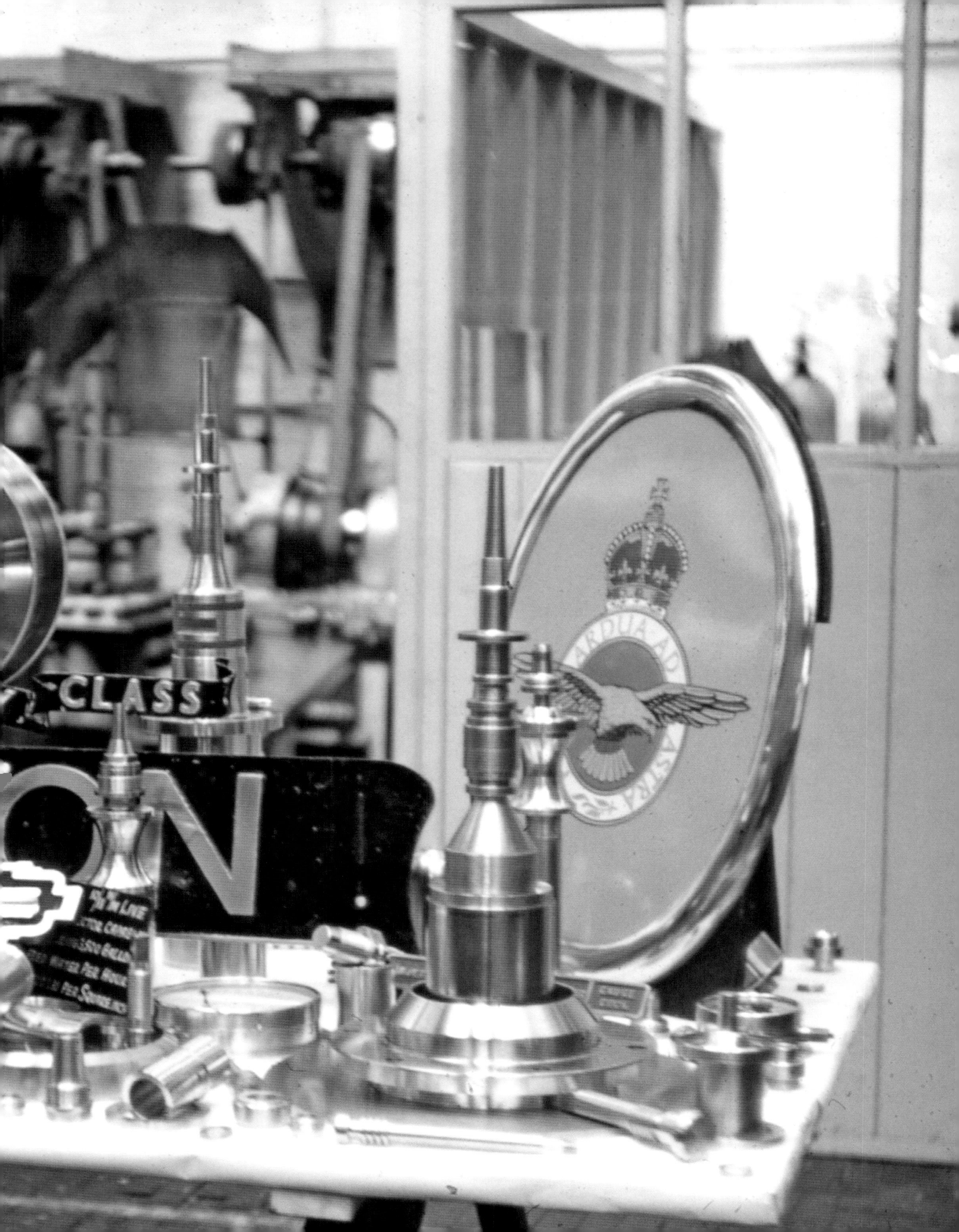
CLASS
ARDUA AD ASTRA

Above: Eastleigh Works in July 1952, No 30854 *Howard of Effingham* is seen nearing the end of repairs; wheels, bogie, dome cover, and chimney yet to be added.

Opposite top: No 70014 *Iron Duke* lifted by the overhead crane in June 1954 and seemingly ready to be lowered back down on to its wheels. Alongside on the next bay is No 35009 *Shaw Savill*.

Opposite bottom: Another 'elevated' view, No 34032 *Camelford* minus front bogie is hoisted clear perhaps ready for its from bogie to be replaced, June 1954? Behind is another Merchant Navy, this time No 35025 *Brocklebank Line*. The cab at ground level took a bit of thinking about but can no be seen to come a B4 0-4-0T.

Next time; S C Townroe's coverage of the 1952 Shawford derailment.

Don't forget, copies of the S C Townroe colour images are available as downloads.

34032

Concluding our works views, the exterior of the works at Ryde St Johns. The view was taken on 18 May 1952 and shows the boiler from an O2 on the left together with No W29 *Alverstone* on the right.

First Generation / Heritage EMUs

I will admit to receiving an amount of angst in the title above when first used in issue No 7. That said I am prepared to stand my ground simply on the basis that none of the EMU designs depicted here remain in service and if it currently fashionable to refer to older infernal (deliberate slip) combustion sets as 'first generation' then why not the same for the Southern sets. So having built some degree of protection here we go again.... .

3-SUB unit No 1758 at Charing Cross. This was one of the final batch of similar units (Nos 1717-1772) which incorporated bodywork from earlier 'Overhead' units mounted on new standard unframes, They entered service in 1929/30. The 'L' headcode was used to indicate either a special working or empty stock. Notice the hinged window on the offside allowing the stencil to be changed and also the position of the whistle by the driver's window.

Above: A 2-BIL, No 2013 at journeys end, Portsmouth Harbour, the service made up of two such units. The 2-BILs were the mainstay of stopping and semi-fast services on the South Western lines for several decades although might also been seen at Reading and on coastal workings, a total of 52 sets built in 1935-37. This particular working was a Brighton to Portsmouth train. Alongside are a pair of COR or similar sets for the main line service to Waterloo. The view is dated c1964.

Opposite top: Alongside a platform where there is a veritable mountain of mail, 2-HAL No 2654 displays a mixture of Southern Railway flair and in the front window placing a more austere Bulleid look. (The term 'HAL' meaning it was 'half lavatory' – one lavatory between the two coaches with no corridor between the pair. The first train were built at Eastleigh and delivered in 1939 intended for use on the Maidstone and Gillingham electrification scheme. In the late 1950s they began to seen elsewhere such as here at Guildford, set No 2654 displaying the numeric code '7' for a stopping service between Waterloo and Portsmouth via Worplesdon, 26 September 1965.

Opposite bottom: Looking east at Blackheath with an augmented SUB unit with the addition of a wider vehicle. One aspect of EMU working that persisted for decades was the use of the oil tail lamp. Electricity would have been far more efficient although perhaps there were reasons why this was not adopted such as a set being stranded on a dead section, 14 May 1955.

6
GUILDFORD
ICE

24

90

Opposite top: The part totem tells us this is Vauxhall and where 4EPB No 5305 is in smart BRS green livery with the coaching stock crest displayed. The '24' number tells us this a Waterloo – Shepperton via Wimbledon train, 13 March 1965.

Oppsoite bottom: A Gillingham to Victoria service diverted (for unstated reasons) at Falconwood, 31 May 1969. Leading is a 2HAP, No 6003. Unusually the train does not appear to display any yellow warning paint at the front – or might be travel worn?

IThis page: Polegate sees 4BUF No 3148 arriving with a Hastings to Victoria working via the Quarry line. The sets incorporated a very modern for the time design of buffet car which proved all too popular with the travelling public for all the wrong reasons. Indeed it was not unknown for passengers to linger over a single cup of coffee for their whole journey and in consequence takings fell. *J Davenport / Transport Treasury*

Treasures from the Bluebell Railway Museum
Tony Hillman

Southern Railway 1931 Census.

The Southern Railway complied a 639 page census of staff in post on 5 March 1931, it was published by the General Manager's Office in September. This listed twenty-two Departments and all their staff under many sub-headings. Staff were also listed by their type of work, thirty-two different headings, which included obvious titles such as 'Station Master' and 'Driver'. More unusual job types were 'Dredging and Tug Boat Staff' and 'Sea-going Staff (Female)'.

The table below shows the staff working on the Lynton & Barnstaple Railway. Given that the Railway closed four years after its publication this could well be the final staffing of that Railway. No Lynton & Barnstaple staff are shown at Barnstaple Town Station so, presumably, the main line staff shown there worked on the Lynton & Barnstaple also.

Station	Department	Staff	
Lynton	Traffic	4	Station Master, Grade1 Porter Signalman, Porter Guard, Senior Checker.
Lynton	Locomotive Running	3	Cleaner, Driver, Fireman
Woody Bay	Traffic	1	Station Master.
Woody Bay	Engineer's	3	Ganger, Underman(2)
Blackmoor	Traffic	2	Station Master.
Blackmoor	Engineer's	3	Ganger, Underman(2)
Bratton Fleming	Traffic	1	Station Master
Bratton Fleming	Engineer's	3	Ganger, Underman(2)
Chelfham	Traffic	2	Grade1 Porter Signalman, Porter Signalman
Chelfham	Engineer's	3	Ganger, Underman(2)
Barnstaple, Pilton Yard	Traffic	8	Carriage Cleaner, Checker, Crossing Keeper *, Goods Porter, Passenger Guard(2), Signalman(2) $
Barnstaple, Pilton Yard	Engineer's	3	Ganger, Underman(2)
Barnstaple, Pilton Yard	Chief Mechanical Engineer	10	Examiner, Foreman, Carpenter, Fitter(2), Fitter's Assistant, Labourer, Painter, Smith, Striker.
Barnstale, Pilton Yard	Locomotive Running	11	Coalman, Cleaner(2), Driver(4), Fireman(4)
Total Staff		**57**	* Braunton Road Crossing. $ Pilton Yard Box

Barnstaple Town in August 1923. The Southern had taken over the L&B from 1 July that year hence the coach remains in original L&B livery. A train is due from Lynton, the coach ready to be attached to the return working in case of need - this obviated the need to move a vehicle from Pilton Yard which was approximately one third mile away. *Bluebell Museum / E. E. Wallis*

Next time in this series: New museum artifacts.... and more from the 1931 census

Special Trains to view the arrival of liners at Southampton

For many years the Southern Railway ran special excursion trains to Southampton Docks when liners were docked there. These excursions were continued by British Railways. Passengers were given conducted tours of the ships. The handbill shows a 'conducted inspection' of the Queen Mary on 30 May 1939.

The RMS Queen Mary was built at John Brown's shipyard on the Clyde. It left the Clyde on 24 March 1936 and arrived at Southampton on 27 March. Unlike previously when just one special train was run to view a liner in dock, numerous special trains were run to Southampton to see the Queen Mary. These trains ran from 27 March until 1 April.

The Museum Archive holds the Southern Railway 13 page document listing the details of the operation.

On Friday 27 March, the day Queen Mary arrived at Southampton, nine special trains ran. The first three ran from Waterloo leaving at 8.20am, 8.25am and 8.38am. Two Restaurant Cars were included in the formation. Each train conveyed 375 passengers and on arrival at Southampton Docks the passengers embarked on the SS Isle of Jersey or SS Lorina to view the liner's arrival. The trains returned to Waterloo by 5pm. The fare of 15/- covered the train and steamer.

Three further trains conveyed a total of 1,000 passengers. These passengers had reserved seats on the seaward facing balconies on sheds number 101 to 106 on the Western Docks. Two trains left Waterloo at 10.20am and 10.34am, while the other left Poole at 11.48am, picking up at all stations to Christchurch. The trains were all Restaurant Cars and returned to the starting point by 4.30pm. The fare from Waterloo was 10/6d and from Bournemouth 5/6d.

A train left Brockenhurst at 8.30am for Lymington Pier for travel by ship to Southampton picking up at Yarmouth, this returned to Brockenhurst at 3.13pm.

Two trains ran to Portsmouth Harbour. One left Hastings at 7.12am, the other from West Croydon departing at 8.9am. These trains picked up at major stations enroute. These passengers boarded the PS Whippingham to view the arrival. A total of 750 passengers were carried. A further departure from Waterloo carried passengers for the SS Twickenham Ferry. This left at 10.45am, included three Restaurant Cars and returned to Waterloo at 6.03pm.

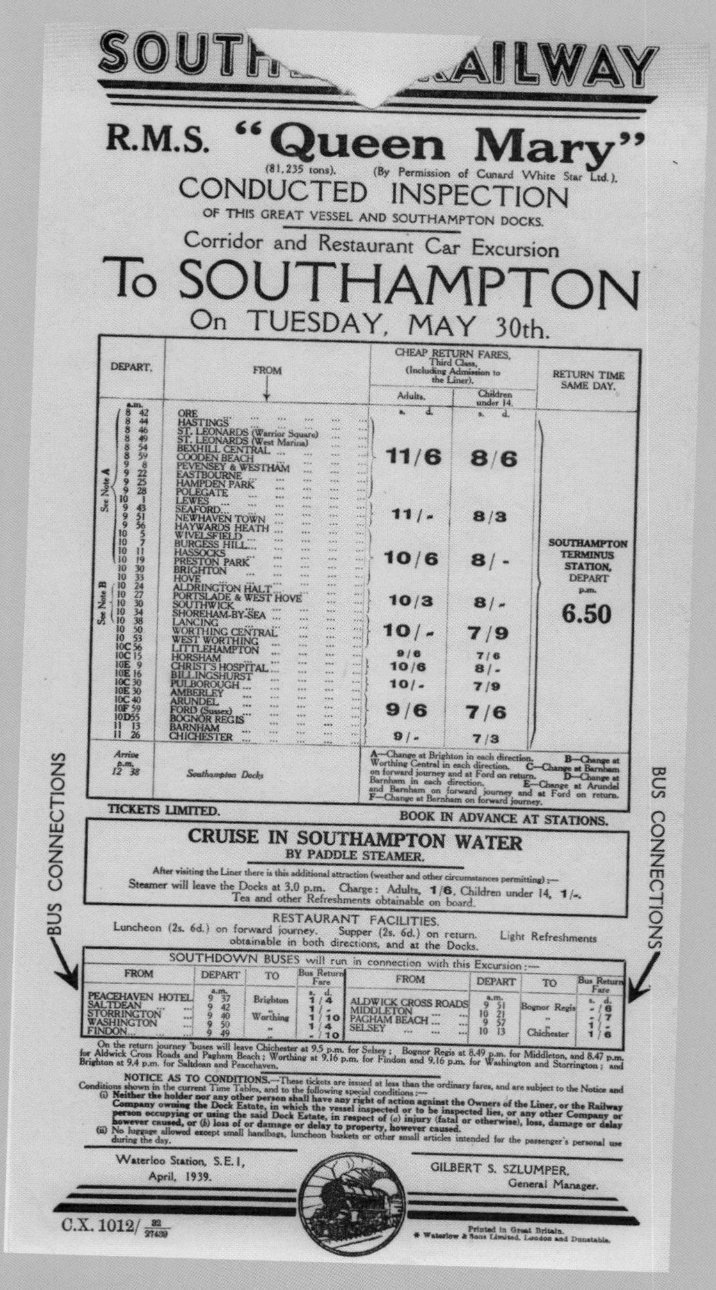

SOUTH[illegible] RAILWAY

R.M.S. "Queen Mary"

(81,235 tons). (By Permission of Cunard White Star Ltd.).

CONDUCTED INSPECTION

OF THIS GREAT VESSEL AND SOUTHAMPTON DOCKS.

Corridor and Restaurant Car Excursion

To SOUTHAMPTON

On TUESDAY, MAY 30th.

	DEPART.	FROM	CHEAP RETURN FARES, Third Class, (Including Admission to the Liner). Adults.	Children under 14.	RETURN TIME SAME DAY.
	a.m.		s. d.	s. d.	
See Note A	8 42	ORE			
	8 44	HASTINGS			
	8 46	ST. LEONARDS (Warrior Square)			
	8 49	ST. LEONARDS (West Marina)			
	8 54	BEXHILL CENTRAL			
	8 59	COODEN BEACH	11/6	8/6	
	9 8	PEVENSEY & WESTHAM			
	9 22	EASTBOURNE			
	9 25	HAMPDEN PARK			
	9 28	POLEGATE			
	10 1	LEWES			
	9 43	SEAFORD			
	9 51	NEWHAVEN TOWN	11/-	8/3	
	9 56	HAYWARDS HEATH			
	10 5	WIVELSFIELD			
	10 7	BURGESS HILL			
	10 11	HASSOCKS			SOUTHAMPTON TERMINUS STATION, DEPART
	10 19	PRESTON PARK	10/6	8/-	
	10 30	BRIGHTON			
	10 33	HOVE			p.m.
See Note B	10 24	ALDRINGTON HALT			
	10 27	PORTSLADE & WEST HOVE	10/3	8/-	6.50
	10 30	SOUTHWICK			
	10 34	SHOREHAM-BY-SEA			
	10 38	LANCING			
	10 50	WORTHING CENTRAL	10/-	7/9	
	10 53	WEST WORTHING			
	10C 56	LITTLEHAMPTON	9/6	7/6	
	10C 15	HORSHAM			
	10E 9	CHRIST'S HOSPITAL	10/6	8/-	
	10E 16	BILLINGSHURST			
	10C 30	PULBOROUGH	10/-	7/9	
	10E 30	AMBERLEY			
	10C 40	ARUNDEL			
	10F 59	FORD (Sussex)	9/6	7/6	
	10D 55	BOGNOR REGIS			
	11 13	BARNHAM			
	11 26	CHICHESTER	9/-	7/3	
	Arrive p.m. 12 38	Southampton Docks	A—Change at Brighton in each direction. B—Change at Worthing Central in each direction. C—Change at Barnham on forward journey and at Ford on return. D—Change at Barnham in each direction. E—Change at Arundel and Barnham on forward journey and at Ford on return. F—Change at Barnham on forward journey.		

TICKETS LIMITED. BOOK IN ADVANCE AT STATIONS.

CRUISE IN SOUTHAMPTON WATER

BY PADDLE STEAMER.

After visiting the Liner there is this additional attraction (weather and other circumstances permitting):—
Steamer will leave the Docks at 3.0 p.m. Charge: Adults, 1/6, Children under 14, 1/-.
Tea and other Refreshments obtainable on board.

RESTAURANT FACILITIES.

Luncheon (2s. 6d.) on forward journey. Supper (2s. 6d.) on return. Light Refreshments obtainable in both directions, and at the Docks.

BUS CONNECTIONS

SOUTHDOWN BUSES will run in connection with this Excursion:—

FROM	DEPART	TO	Bus Return Fare	FROM	DEPART	TO	Bus Return Fare
	a.m.		s. d.		a.m.		s. d.
PEACEHAVEN HOTEL	9 37	Brighton	1/4	ALDWICK CROSS ROADS	9 51	Bognor Regis	-/6
SALTDEAN	9 42	"	1/-	MIDDLETON	10 21	"	-/7
STORRINGTON	9 40	Worthing	1/10	PAGHAM BEACH	9 57	"	1/-
WASHINGTON	9 50	"	1/4	SELSEY	10 13	Chichester	1/6
FINDON	9 49	"	-/10				

On the return journey 'buses will leave Chichester at 9.5 p.m. for Selsey; Bognor Regis at 8.49 p.m. for Middleton, and 8.47 p.m. for Aldwick Cross Roads and Pagham Beach; Worthing at 9.16 p.m. for Findon and 9.16 p.m. for Washington and Storrington; and Brighton at 9.4 p.m. for Saltdean and Peacehaven.

NOTICE AS TO CONDITIONS.—These tickets are issued at less than the ordinary fares, and are subject to the Notice and Conditions shown in the current Time Tables, and to the following special conditions:—
(i) **Neither the holder nor any other person shall have any right of action against the Owners of the Liner, or the Railway Company owning the Dock Estate, in which the vessel inspected or to be inspected lies, or any other Company or person occupying or using the said Dock Estate, in respect of (a) injury (fatal or otherwise), loss, damage or delay however caused, or (b) loss of or damage or delay to property, however caused.**
(ii) No luggage allowed except small handbags, luncheon baskets or other small articles intended for the passenger's personal use during the day.

Waterloo Station, S.E.1,
April, 1939.

GILBERT S. SZLUMPER,
General Manager.

C.X. 1012/ 82/27439

Printed in Great Britain.
Waterlow & Sons Limited, London and Dunstable.

A Steamer Excursion ran from Ryde Pier Head with bookings taken at Isle of Wight stations.

Following arrival, Queen Mary went to the King George V Dry Dock during the afternoon to complete the internal fitting. Some of John Brown's staff went to Southampton to do the work.

On Sunday 29 March, eighteen trains were due to run to view the liner in the Dry Dock. While the trains on Friday all originated on the Southern, many of the Sunday trains came from further afield. Seven came from the Southern, three from Waterloo and one each from Brighton, Sandwich and Portland with a spare slot if required. Six arrived from the Great Western, originating at Acton, Slough & Reading,

Bristol, Hereford, Swansea and Wolverhampton & Birmingham. Five originated on the LMS, arriving from Derby, Harpenden, Leicester, Bletchley and Stoke. Special trains continued to run on Monday, Tuesday and Wednesday the following week.

On those three days seven trains were due to run. Four from Waterloo, one each from West Croydon, Windsor and Exmouth.

Queen Mary left Southampton for sea trials around the Isle of Arran in April before returning to Southampton for the maiden voyage to New York on 27 May.

Queen Mary leaves Southampton on 15 October 1953. ***Photo from Bluebell Museum Archive, Colin Hogg collection***

Remembrance class nameplate and plaque.

A footnote to the article on the N15Xs in Issue 7 of 'Southern Times'. The Museum at Sheffield Park has on display the nameplate Stroudley from the N15X version of the engine plus the plaque carried by both the original Baltic Tank and the N15X. To complete the set the plaque is displayed with the original nameplate Remembrance from the Baltic Tank. We are very fortunate to have the original set as they were spilt when the plaque was transferred to the N15X. However, they are now back together in the Museum.

Dick Hardy at Stewarts Lane

When one considers the number of sheds - and by this we include EMU depots - on the Southern, the years they have existed and the number of men who worked from and therin, the numbers of individuals involved is likely staggering. Be assured staright away this is not an attempt to list same, just to pass comment that for the years and individuals involved, very little has been written based on first hand accounts.

We are of course grateful for that which has survived, a rare example being that fromn Richard Harry Norman 'Dick' Hardy whose recollections as Shed Master at Stewarts Lane in the early 1950s were published by Ian Allan in 1971 as part of a wider recollection of his railway memories.

Fortunately Dick Hardy was also handy with his camera and his captions, his railway collection having survived and we are delighted to be able to reproduce a selection of views from that time having the added benefit of his original captions.

May 1959 at Victoria. This was the day that I travelled on the footplate from Victoria to Paris and worked my passage throughout; as driver to Dover, and fireman Calais-Paris on 231 E26; Mécanicien Bebert Bethune, Chauffeur René Sene and a good bottle of red wine!

In the photo are Sammy Gingell (standing left), former Stewarts Lane driver, retired and now an outside porter at Victoria, a job he did for the next twelve years and which he greatly enjoyed.

Next to him is my great friend James Colyer-Fergusson who made these trips to France possible, then Personal Assistant to the Chairman, Sir Brian Robertson. Driver Syd Patrick of Stewarts Lane and Fireman Brian Matthews, seen here as a top link fireman but just a fireman when I left the 'Lane four years and five months before. How times changed, for Syd was firing at Nine Elms for twenty four years before coming to the Lane before the war as a driver.

Despite the detail the one thing missing was the identity of the engine!

All images, Transport Treasury.

Above: No 34088 *213 Squadron* prepared and about to leave the shed for Eastleigh to work the Royal train from up from Portsmouth Harbour to Waterloo via the Mid-sussex line next day. This was the occasion of the state visit of Emperor Haile Selassie of Abysinia visiting this country in Oct 1954 as the guest of Her Majesty The Queen.

Left: The final touches before departure light to Eastleigh for the special working: prominent are the Royal Buffers and the Royal drag hook and screw coupling. Next day, the Royal brass beaded and polished disc boards will be set in the usual code for Royal trains. Staning left is Bill Thorburn, an outstanding chargeman cleaner who worked extremely hard and led from the front. He kept his Battersea boys in order and there was never a dirty engine on his shift. I inherited two other chargeman cleaners and moved both as soon as possible to a different class of work so that our engines, for a London shed, and remembering that we had no adult engine cleaners as did Kings Cross, were kept in pretty good order. Chief Inspector (standing right) Danny Knight was in charge of the Inspectorate at T E Chrimes' HQ(MPS) at Waterloo and who rode on all Royal or Deepdene duties. He had also selected the two sets of Nine Elms enginemen (Swain/Hooker and James/Reynolds) for the 1948 interchange trials and was liked throughout the SR.

Opposite top: The cab of No 30770 *Sir Prianius* in Aug 1953 and the best I could do with a box camera. The low roof, round-topped firebox and high footplate combine to make the enginemen's world confined but not difficult to work in. The driver is Sammy Gingell, wearing a beret, a rarity on the SR, but Sam liked to cut a dash in a quiet way. His mate is Les Penfold and they must have been together two years in the Ramsgate link before Les moved over to the 'Brick' (Bricklayer's Arms) in, their Dual Link. A typical Eastleigh footplate with Drummond fittings. The firehole door goes back to Stroudley's days on the LB&SC and to the LSWR via Drummond, his Works Manager. The gauge glass protectors are unique to Eastleigh with a spiked column on the right. It was said that LSW firemen kept the water level with the top of the spike. Three steam valves in the centre are for steam heat, supplementary steam for the F class exhaust injector and for the sight feed lubricator behind Les's left shoulder. High up are the steam valves for the two injectors.

Bottom: Taken at Bournemouth West. The serious gent on the left is Major Harry Mosse who, with Bert Hooker, became dear friends. Harry loved nothing more than a day with steam and it was he who thrust his way into my office at Liverpool Street in 1959 and asked very directly whether he could have a journey to Norwich on a 'Britannia'. We 'changed' his regiment from the Gunners to the REs and gave him the necessary and he never looked back. Bert Hooker had vast knowledge and to listen to him was an education. The fireman's name has gone from my memory but never the other two. I gave the address from the pulpit in Mendham Parish Church at Harry's funeral and at Bert's in Eltham Crematorium. Bert began at New Cross Gate in 1934 aged 18 and transferred to Nine Elms as a fireman in 1940. After that he never looked back and, when he retired, he had done it all.

34059

3037

Opposite top: Bulleid on the Eastern...! 'Spl 7' (on the headcode disc) my foot! Here is 34059 *Sir Archibald Sinclair* at Parkeston Quay in 1949 when she was on trial. The Railway Executive intended to transfer what they considered to be under-used light Pacifics from the West Country to the Great Eastern section based at Stratford and Norwich. Meanwhile our Chief, L P Parker, was party to plans for an entirely new high speed timetable using what he hoped would be the class 7 BR standard design then on the drawing board. In no circumstances would he have any Bulleids at Stratford on a permanent basis although the engine did distinguish itself on trial with a picked crew and Chief Inspector Len Theobald in attendance. This was the penultimate day of the trials and we took a very heavy train to Parkeston with great ease in about 85 minutes. Left to right are: HQ Inspector, Tom Sands, ex M&GN who retired as Chief Inspector, Norwich; Dr Bill Burritt of Stratford and his regular mate who later moved to Gorton; Len Theobald, an outstanding Chief Inspector for whom L P Parker had great respect. In preparation we had the tender emptied the tender by Stratford to load it with 'good stuff' for the high speed trial to Norwich the next day. On the return near Swainsthorpe the steam pipe to the steam reverser split and we made the entire journey in full forward gear unable to alter the setting.

Opposite bottom: Reginald Jennings ('RAUJ' or 'Jumbo' to Marlborough College boys) was a famous Housemaster in his prime both before, during and after the last war. He had always wanted a main line trip on steam but for years he could not find the time when I could have arranged it. But in 1966, the deed was done and we took him from Basingstoke to Bournemouth Central on a Merchant Navy and then back to Southampton on No 73037: at the time a Western Region engine and not a bad old boat either. We had prevailed on Reginald to have a little drive which he did up the bank from the New Milton start and onwards to Brockenhurst where the photo was taken.

Above: A very early introduction to the steam locomotive.Bert Hooker and a young Peter Hardy, born in 1958. The engine is No 73082, one Sunday morning at Waterloo.

Note - all images are scanned to the edges of the negative.

7P5FA

The South Eastern from Ashford to Dover and coastal problems for William Cubitt

Jeremy Clarke

Having got over the disappointment of Parliament backtracking on the approval of its independent line to Dover in favour of sharing 11½ miles of the Brighton line north of Redhill, and the time wasted investigating John Rennie's ultimately abortive Central Kent Railway of 1838, the South Eastern's engineer, Sir William Cubitt, managed to get his line open as far as Ashford by 1 December 1842. But conquering the final ten of the next twenty miles to the Channel port was to prove much more demanding.

Eastward from Ashford, which is just over 56 miles from Charing Cross via Sevenoaks, the basic geological stratum is made up of the well-drained sandy soil of the Hythe and Folkestone Beds that run below the scarp face of the Downs. It was when he got into Folkestone Cubitt's problems really began. He was first faced by a deep, steep-sided valley over the Foord river as it forces its way to the coast through a band of gault clay. The treacherous gault also underlies the porous layer of chalk forming Folkestone Warren and itself being impervious is the cause of the instability of this area. It continues to give reason for watchfulness over the line despite a system of preventive measures being undertaken and introduced particularly post-nationalisation. Beyond The Warren Cubitt had to deal with the three miles through the chalk of the North Downs as they turn toward the coast to form the White Cliffs. Not only did the latter material require some careful tunnelling it was also the scene of the wholesale destruction of a promontory which, in 21st century eyes, would certainly be viewed as nothing less than an officially-sanctioned ecological and environmental disaster.

Those twenty miles between Ashford and Dover resemble the end gable of a house as the line climbs for some 8½ miles through and out of the valley of the Great Stour to a summit about 240' above mean sea level just east of Westenhanger station, and then falls back to sea level at Dover's Archcliffe Junction. That makes an average gradient of 1 in 266, a trying start for a 'cold' engine with a heavy London-bound boat train. The eastward climb from Ashford is not constant, for it contains a spot of 'level' and one very short dip, though the final four miles are unbroken at the relatively easy gradients – by SER standards – of 1 in 250/280.

The line out of Ashford is level for the first half-mile or so as the junctions with the Hastings and Ramsgate routes are negotiated as well as the connection to the Hi-Speed line that has passed by the station on the down side. That part of the line flies over the Ramsgate route and then parallels the main line until bearing away to the Channel Tunnel terminal at Cheriton, west of Folkestone. In the course of this it passes over part of the site of Ashford loco shed which was tucked into the triangle formed by the Main and Ramsgate lines with access from the latter. Its original site was to the south of the main line next to the works - which lay at 'E' junction where the Ramsgate line turned away to take the Stour valley northward - having opened there when the SER reached Ashford from Headcorn in November 1842. The shed of about 200' long housed four 'through' roads leading to a neck at the east end, the coal stage and turntable being in the western approach.

The Southern planned the later shed in 1927 and opened it in 1931, the layout being in similar style to those of the sheds of slightly later date at Hither Green and Norwood Junction. The building housed ten roads under an eight-bay northlight roof. Two of the roads passed right through the shed to a neck at the extreme east end of the site. The turntable was set in the western approach while the raised coal stage lay alongside the north eastern boundary. The shed provided motive power for a variety of services though these were mainly of a local nature. At Nationalisation

Opposite top: In the cab of No 34101 *Hartland* ready for the 11.00 Victoria - Dover boat train is none other than W O Bentley. 'WO' is is relaxed and happy as well he might be having been a premium apprenticeship with the Great Northern from 1905 - 1910; how he loved the return to the railway that I arranged for him after we first met in November 1958. He went many miles on the footplate until his last trip in 1961 with Driver Percy Tutt and Fireman John Hewing - both seen here on the platform. What a wonderful man. His was a hard life but his achievements were many and great to say the least. We had the same engine for the retrun working from Dover, which was very hard work indeed for the fireman, who was me!

Opposite bottom: No 70004 *William Shakepeare* has long gone and what remains of the Golden Arrow stands in No 8 at Victoria (E). The train now goes once more to Dover Marine leaving London at 10.30. The Dover men shown have no doubt come up with the 'Blue' (The Night Ferry) and probably with the same electric locomotive, No E5000. These machines were master of the Night Ferry although they had some trouble, I believe, with the gaps on the climb up to Sheperdswell from Dover.

there were sixty engines on the books with 0-4-4Ts and 4-4-0s predominating , although eight 'N's and four 'N15's from the last batch built with six-wheel tenders originally for Brighton services were also allocated for the heavier work. Following electrification it closed to steam in 1963, diesel servicing finishing five years later when a Steam Centre was established on the site. This closed in May 1976, the area later being cleared and recently sold for housing development.

On leaving Ashford the Dover line follows the Stour's upper reaches for nearly eight miles with 1¼ miles of that distance up at 1 in 260 in which it passed Sevington box, 1¾ miles from Ashford. This tiny timber structure on the down side controlled access to a public siding as well as overlooking an occupation crossing. It also conveniently broke up the 4¼ mile-long section to Smeeth, but closed with the commissioning of Ashford panel box in April 1962.

A half-mile long level section beyond Sevington preceded continuation of the climb before the line dipped down at 1 in 225 to Smeeth, which opened in October 1852, nine years after the line at 60m 33ch from Charing Cross. This was to the typical SER staggered platform layout, the up platform being to the west with the station building and weather-boarded signal box at its down end. A single siding at the country end on the up side catered for freight traffic. The building was unusual in being multi-gabled though the gable at the up end was a later addition. The small and scattered community in the vicinity could not sustain the station's use post-nationalisation, losing its passenger service early in 1954 though freight traffic survived for another ten years: the signalbox closed in April 1962.

The short downhill run from Smeeth is succeeded by the 1 in 250/280 climb towards Westenhanger. A little less than halfway up this Herringe box once stood just short of the 62¼ milepost. This small weather-boarded building was not in regular use, being opened only at busy times. It gained rather more importance in

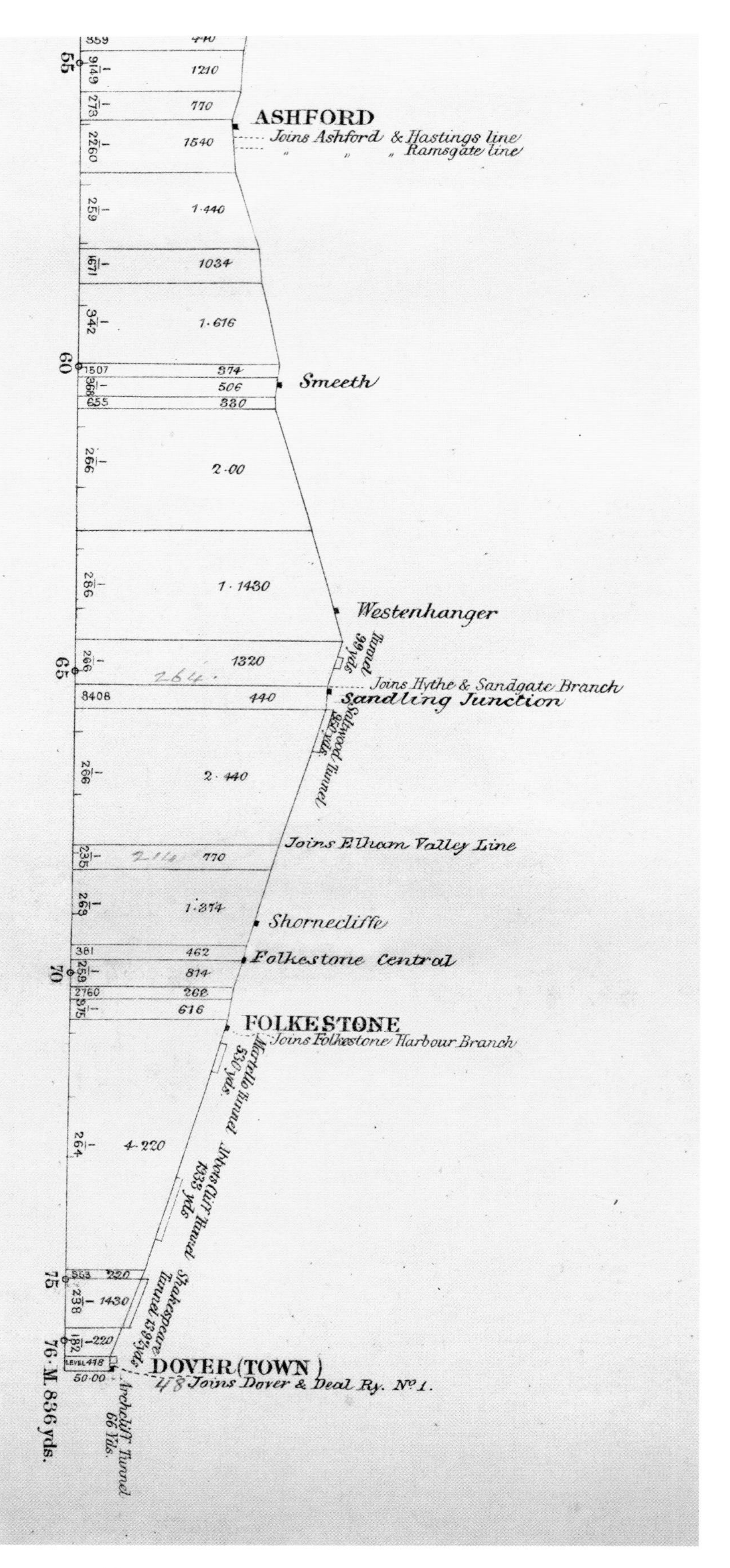

Leaving a trail of exhaust in its wake, No 34031 *Torrington* passes Ashford c1948/49. The two discs inditate a train from Victoria. *FH / Transport Treasury*

World War II when two sidings were laid on the up side of the line for rail-mounted guns though the track had been lifted soon after nationalisation.

Westenhanger Racecourse station opened in 1905 as an extension of its progenitor. Situated a quarter-mile to the west of it and right beside the racecourse lying on the up side, it consisted of four platform faces made up of two islands, its use being confined to race days. Despite closing in the 1960s remnants of the platforms remain.

Westenhanger station itself is just over eight miles from Ashford and opened with an '& Hythe' suffix six months after the line, the only intermediate stopping place at the time between Ashford and Folkestone. Its platforms stand either side of the bridge carrying Stone Street over the route, a former Roman Road linking Canterbury and Lympne. The up platform, to the west, had a short up bay behind it and bears the two-storey brick station building constructed in 1861 in place of the original timber structure: it underwent restoration in 2009. A down bay ran almost the full length of the main down side platform. The goods yard was also on the down side opposite the up platform and consisted of little more than a single dead end siding alongside a loading dock and the small brick warehouse – the term generally used by the SECR rather than 'goods shed'. The signal box – three different ones over a period of time – was beyond this in the shadow of the road bridge. Proposals for closure of the station were aired in 1969 but rejected though it has been unmanned since then.

A little more than a quarter-mile beyond Westenhanger the line reaches its summit and begins the eleven-mile run down to the sea at Dover. Having passed through the 100 yards of Sandling tunnel the route comes to Sandling at 65½ miles from Charing Cross. It opened as Sandling Junction in January 1888, being the point at which the double-track branch to Hythe and Sandgate left the main line. Two signalboxes governed movements, no 2 box at the down end of the branch up platform, no 1 at the junction. This closed with the introduction of colour-light signalling early in 1962. Until Sandling's opening Westenhanger had acted as the junction station though it lost its 'Hythe' suffix when the branch got that far in October 1874. This was the anticipated first stage of a line directly to the port facilities at Folkestone, making a much more measured approach than the steep and costly-to-work branch from Folkestone Junction with its inconvenient shunting moves. The SER Chairman, Sir Edward Watkin, had been elected MP for Hythe in 1874 and perhaps as a means of maintaining the favour and approval of his constituents later had to concede that a double-track main line running through it met locally with considerable influential resistance. What could likely have been achieved with little difficulty in the 1840s proved impossible thirty years later.

SANDLING JUNCTION

Opposite top: Sandling Junction viewed towards Folkestone. An unidentitified 'Lord Nelson' is passing through on an Up express, c 1954. *MC / Transport Treasury.*

Opposite bottom: This time a Down train behind No 30928 *Stowe* on 25 August 1951. The Hythe branch service behind an 'H' class tank is at the branch platform, services on the branch ceased just three months later. *R C Riley / Transport Treasury*

Top: Nine years later and the track leading on to the branch is still present as are a pair of Pullman Camping Coaches. The third rail has been added to the main line but for the present semaphore signals remain. *PY / Transport Treasury*

Bottom: later still, the stop signal has been replaced with three aspect MAS, again looking towards Folkestone. *Transport Treasury.*

Folkestone Warren 1877 when some 60,000 tons of chalk fell. It took two months to restore the railway to its original state.

Sandling's single-storey station building, in brick and timber with a steep, gabled tile roof, stood on the up side of the Hythe branch and thus became remote from the platform when the up line here became a goods loop with the singling of the branch by the Southern in 1931. The branch closed entirely in December 1951 at which time Sandling lost its 'Junction' suffix. The building lasted for some years though it has since been demolished and its responsibilities taken over by the former waiting room on the main line up platform.

A level section of about ¼-mile follows before the downward gradient resumes at 1 in 270 through the 954 yards of Saltwood tunnel and, 1¾ miles beyond Sandling, the line passes the site of Cheriton Junction. It was here that the Elham Valley Railway completed its sixteen mile-long journey from Canterbury. This line had been promoted locally in 1861 as a double-track railway but as neither the SER nor the LCDR was prepared to work it the scheme failed. But when the Chatham proposed to link Canterbury and Folkestone an alarmed South Eastern took it up but as a rather lighter railway than the original: authorisation came in July 1884. The ten miles northward from Cheriton to Barham opened three years later and the remainder to meet the Ashford-Ramsgate line south of Canterbury [West] station in 1890.

The army took over the railway during the First World War and introduced single line working because much of the track was used for storage. Following the massive slip in The Warren in 1915 the EVR also had limited use as a diversionary route. It reopened in 1919 but from 1931 the Southern worked it single line only over the thirteen miles north of Lyminge. World War II saw the military take over again and close it to all public services. The cost of refurbishing the line at de-requisitioning could not be justified and though some freight traffic ran the Southern closed it completely in 1947. Lifting was a desultory process carried out between 1950 and 1954.

Cheriton Halt stood a half-mile east of Cheriton Junction and was served only by Elham Valley trains. It opened on 1 May 1908 but closed in December 1915. Interwar use was from July 1920 to February 1941: it opened post-war for a period of just seven months, closing on 16 June 1947.

Folkestone West station opened on 1st November 1863 as Shorncliffe Camp, 69¼ miles from Charing Cross. It was rebuilt and enlarged in 1881, quadruple track passing through it in the usual SER fashion of main lines between platform loops. An additional line from Cheriton Junction for Elham Valley traffic met the down loop at its up end but also provided access to an up bay with runround facilities. The runround had a siding off it into the generating station of the local electricity company.

A small dock lay at an angle at the down end of the up platform while the goods yard, of five sidings of various lengths, was behind that same platform. Further west still a siding led into a storage area belonging to the local authority. The 'No.1' signalbox was also on the up side at the end of the dock, No 2 being at the down end of the up platform: it closed in 1930. The station lost its 'Camp' suffix in 1926 and its main goods facilities in April 1965 though coal traffic continued for another three years.

The Kent Coast electrification scheme saw the four track layout at Shorncliffe extended right through from Cheriton Junction to a point just west of Folkestone Central in 1961, though only double track survives now. Shorncliffe was renamed Folkestone West at the same time.

The most impressive – if only single-storey - brick building appeared on the down side of the station at the 1881 reconstruction. With its arched and paired windows, massive chimneys and the prominent ends to the roof rafters it lends its style to the later buildings at Chislehurst and Elmstead Woods dating from the widening of the Sevenoaks line out to Orpington soon after formation of the SECR in 1899. The up side building of the same style and almost the same size faces it, both buildings still in being.

Still travelling steadily downhill the line reaches Folkestone Central at just under seventy miles from Charing Cross and 13¾ miles from Ashford. This opened as a temporary terminus in 1843 but it was to be another forty years before a permanent through station was completed, then named Cheriton Arch. This was to the west of the main A259 coast road - named Cheriton Road here - which passes under the line, the terminus having been on the east side. The platforms were reached by covered ways leading up from the road, the main building being on the down side. A long up bay nestled up to the buildings sited at that platform's down end while the small timber signalbox lay off the London end of the down platform. The station was renamed Radnor Park in 1886 and Folkestone Central on 1st June 1895.

Electrification saw the station completely rebuilt with two island platforms linked by a subway from the booking office, steel canopies and luggage lifts. The up side island was abandoned in 1999 when the four tracks to Cheriton Junction were reduced to two and though the platform remains all the buildings have been removed.

A little less than a half-mile further east the line reaches the nineteen arches of the magnificent Foord viaduct standing up to 100' high as it strides across the now-residentially covered river valley. In recent years the arches have been strengthened with steel tie rods.

A further half-mile on and a little less than fifteen miles from Ashford comes Folkestone East, which opened with the line in 1843 as plain 'Folkstone' – note without the middle 'e'. The 'Junction' suffix was added in 1849 with several variations on the theme, including 'Folkestone Old', before its final identity - Folkestone East – in September 1962.

The station at first consisted of the usual staggered platform arrangement and was the nearest to the harbour then beginning development by the South Eastern. But principally it was the junction with the one-mile long branch down to the Harbour which also opened in 1843, though for goods traffic only. The Board of Trade refused sanction for passenger use until 1849: until then travellers to and from the Continent had to find their own way between the main line station and the steamers. When passenger services over its 1 in 30 gradient did begin the branch was then identified as the steepest in the country regularly used by main line trains, albeit with rather less glamorous motive power than brought them there or took them away.

The SER's main freight facilities for the town were concentrated at the junction. By the end of the nineteenth century a substantial goods and coal depot had been developed within the angle of the main line and the departing harbour branch though the goods shed and cattle pens were to the north. Thirty years on the goods facilities had been much expanded, the goods warehouse and cattle pens having been moved south into this area too, while a three-road engine shed and 65' turntable had appeared to the north opposite the branch junction. The great majority of traffic on the branch emanated from or went to London so the fact the junction faced east made the transitions unnecessarily involved. Passenger traffic was brought into one of two long sidings on the up side for removal and coupling up of the appropriate engine(s). That area is now occupied by a half-dozen berthing sidings each fourteen coaches long, three of which are only accessible from a stub of the branch line.

The small brick and timber signalbox lay at the end of the down platform opposite the first loco depot but this was superseded soon after the turn of the 19th century by a much more substantial one on more or less the same site. This was classified 'A' up until 1930 when Folkestone 'B' box, actually about half-mile away beyond Martello tunnel, was closed. Until its closure in 2018 following extension of Ashford's control the 'A' box assumed responsibility for the line east of Folkestone Central to Buckland Junction and from there to meet Shepherds Well on the ex-LCDR Dover route and Walmer on the Deal 'joint' line.

Following the opening particularly of the reconstructed Folkestone Central passenger traffic to Folkestone East went into slow decline. Despite the expenditure on rebuilding the station in 1960 with the onset of Phase 2 of the Kent electrification scheme, it closed on 6 September 1965. Little remains other than short lengths of the down platform for staff purposes as does much of the up one for the same reason, neither now graced by any buildings.

Electric services over the Harbour branch began in June 1961, the ex-GWR 'panniers' that had taken over banking duties from the venerable 'R1' tanks losing out to the new EMU's. Withdrawal of freight traffic over the line occurred in August 1968 and the few remaining regular passenger services went in 2001. However, VSOE trains continued to use the Harbour station until 2008 and a steam-hauled rail tour travelled the line in March 2009. (VSOE passengers now alight at Folkestone West and are taken by road to the Eurotunnel terminus at Cheriton.) The branch formally closed at the end of May 2014, the iconic viaduct and swing bridge at the harbour having been Grade II listed by English Heritage in January 2012. Much of the Harbour station survives and at the time of writing a degree of development is being considered: a footpath now traverses the length of the branch itself.

The 533 yards of Martello tunnel - named after the tower that stands above it – are driven through gault clay, greensand and chalk, making it technically the most difficult of the once-four tunnels between Folkestone and Dover. But there follows another major difficulty facing Cubitt, one he could not tame, Folkestone Warren. The first 'slip' actually to be recorded here took place in 1765. A movement of such severity as to possibly threaten the line's future occurred in 1877. It took two months to clear the estimated 60,000 tons of material, rebuild over one hundred yards of Martello tunnel and relay the track across the slip. The SER's Chairman, Sir Edward Watkin, accused the LCDR Board of starting and spreading rumours the damage was so severe the line would never reopen, to the detriment of the SER's share price. On the other hand it was recorded that on hearing news of the slip the LCDR Chairman, James Staats Forbes, offered use of the Chatham line to the South Eastern. That offer seems unlikely but knowing Forbes' reputation it is quite possible he sowed the seed of the idea and let it grow.

Five years later another serious fall buried the line west of Abbotscliffe tunnel under twenty feet of chalk, one of the eight slips recorded before the turn of the nineteenth century. Despite the damage these slips caused it could not compare with that left by the massive landslide on 19 December 1915 when some 1.9 million cubic yards of chalk over a distance of about one mile slipped up to fifty yards towards the sea, marooning a train that had just left Martello tunnel. Collapsing cliffs buried parts of the line to a depth of sixty feet and in places pushed out material across the beach and into the sea for up to eighty yards. Fortunately nobody in the train was hurt and the passengers and crew were able to return along the beach on foot to Folkestone Junction.

To be continued in Southern Times No 9

From the Footplate

We start with letter from **Richard Vidler**. 'Thought I would get in touch, in case nobody else does, about the **photograph on page 48 of Issue 6.**

'The photograph is actually a bit more interesting than you might have imagined, I actually picked up a b/w negative (via a certain interne auction site) a little while back, it came with full details of date and location.

'It is Chipstead on the Tattenham Corner branch and was taken on Wednesday 31 May 1961. They would stable a loco in the yard there on Derby Day in case the Royal Train loco had problems.'

Now from **George Hobbs re ST7.** 'Another great read, thank you. A couple of minor points. On page 62 the unidentified 4-SUB is entering platform 1 at Charing Cross, almost certainly a Dartford Loop line service. The top picture on page 70 is just outside Cannon Street. Not sure how both of these were ascribed to be London Bridge.'

Next from G**raham Buxton-Smither.** 'Dear Kevin, I hope you can forgive me for asking a question which relates to your time wearing your previous editor's hat but you had planned to publish it in a future issue (55?) but left Southern Way beforehand; I'd still love to know the answer...
The following is an extract from the message originally sent to you and contains the relevant details:

'Does the Duke of Norfolk have any special rights concerning the services through Arundel Station? An odd question you may correctly surmise but it was the memory of a very surly driver that brought back a memory.

' The late 17th Duke of Norfolk, Miles Fitzalan-Howard, was a chum who I sometimes dined at my old club, the Naval & Military, and on the last occasion we met before his death in 2002, I joined him for a convivial lunch at Arundel Castle. It was one of those happy days I used to enjoy when I was still working in London as I

K No 32340 standby Royal engine at Chipstead, 31 May 1961. ***Richard Vidler collection.***

had another engagement that evening at the Chelsea Physic Garden for an opera picnic to raise funds for a children's medical charity. I confided to Lady Georgina, Miles' daughter in law, that I had to catch a specific train or I'd be in real trouble as I was to be in the company of Prince Edward and Loyd Grossman and so should not arrive after the Royal guest. The luncheon became ever more convivial until it got to the point that Georgina, remembering my need for the train, exclaimed that it was due to depart the station in 3 minutes. Horrified, I ran for the Gate at which Miles' had kindly arranged a taxi, shouting as I left my companions, 'Tell them to hold the train – I'm on my way...".

'The taxi sped down the hill, through the town, over the roundabout and into the station forecourt which fortunately was by the up-line. The platform entrance was near the front of the train and I ran through seeing that there was still a train on the platform and the small First Class section was right opposite. As I reached the train, grabbing the door-handle, a very surly driver, hanging out of his window looked at me and said something along the lines of: 'I suppose you're the **~!~~**! bloke that we've had to wait for?' and his face disappeared from view with those words ringing in my ears. Someone closed the door behind me and off we went. Thankfully, the return journey was otherwise uneventful and we arrived in London on time.

'I just took it for granted that Miles had told the train to wait but after more sober reflection, I now wonder if he had merely prevailed on the station staff to do him a kindness and let me on. Hence my question; I'm aware that some landowners have certain rights over services operating on their land and wondered if the Dukes of Norfolk have any such entitlements at Arundel Station?'

Now from **Alan Holmewood on ST7**. 'Once again a fascinating issue. One or two things come to mind. Farnborough Main signalling. In the section dealing with the 1947 collision, I think the substation involved was at Sturt Lane, rather than Stud Lane' – Ed. 100% agree. It was written correctly in the draft and somehow I suspect 'AI' changed it!). 'Kemp Town. As Roedean School is a girls-only establishment, I doubt that the youngsters wearing boaters came from there!

'First Generation EMUs. I have to query the caption on page 62. The whole surroundings do not square with London Bridge, and that station was very much "up side and down side", so an arriving train would not be on that side of the formation. From the bridge girders, I suggest that the train is arriving in Platform1 at Charing Cross. I refer you to https://www.wbsframe.mste.co.uk/public/Charing_Cross.html which I think supports this. This simplifies the headcode question; David Brown's work lists this as Gillingham/Gravesend/Dartford etc to Charing Cross via Sidcup and Lewisham. The same code would have beeen used on any service via this route from any of the Eastern Section termini. Different codes would have been used for up services destined for Cannon Street or Holborn Viaduct.

'Uckfield. Some clarification re the photo on page 80. The signalbox remains and is in use as a taxi office. The original station has disappeared and is replaced by a single platform just on the London side of the box. There is a photo of the box here: https://commons.wikimedia.org/wiki/File:2011_at_Uckfield_station_-_signal_box.JPG '.

Paul Cooper who has much personal knowledge of the Farnborough area, was able to afford much additional detail to the article in **ST7**.

'Southern Times. Lovely edition. A few comments on Signalling at Farnborough. Note on P.29 you can see the gantry on the up roads at the end of the platform. But in the distance you can see some railway relics of the station names of Shipton on Cherwell and Horsmonden. These plus others (including the Devon Belle deflector plates) all belonged to the Railway Enthusiasts Club that was based in a hut on the site the other side of the Portsmouth Road bridge. It was there until the early 70's when the site was redeveloped. It continues to function in Farnborough on a different site.

'P33 states the gantry illustrated is the first west of Farnborough. It is almost certainly the second one towards the old Bramshot Halt between Farnborough and Fleet. I spent many a day either side of the gantry taking photos 1963-6 and recognise the fields to the right and the. In fact for several weeks at the end of 1966 the old signal arms were simply thrown to one side. Whilst waiting to photo the last knockings of steam, I liberated the only intact distant signal arm to the other side of the fence. I only had a pushbike at the time so covered it with foliage from the extensive ferns by the side of the fence. Over two years later, by which time I had a minivan, I re-visited the site to find the arm intact, still under even more accumulated foliage, and liberated it to my rented flat. It followed me around for the next 50 years ending up in my garage with my kids using it to lock their bikes against (the spectacle plates were broken by this time). The arm was still unrestored and had the patina of plenty of smoke protecting the original paint. In the end it went to a good home some 5-6 years ago.

'Now the other point relevant to this, is that if you turn

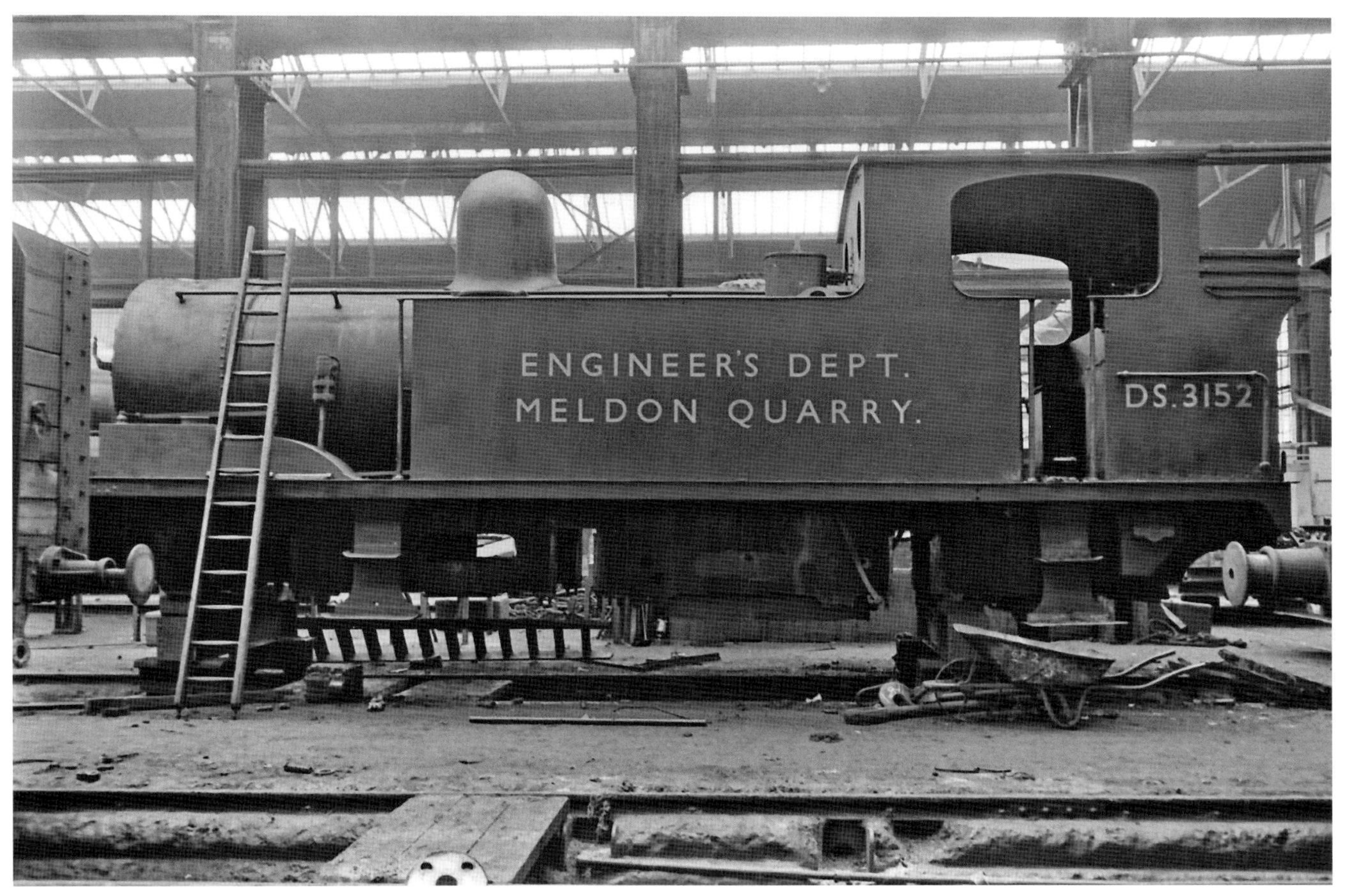

Dick Henrywood kindly send this view of G6 No DS3152 receiving attention to its lower parts – probably recorded at Exmouth junction. Dick is specially looking for images of Bow station. We have advised a few sources but any additional ideas would be welcome.

to P2, this almost certainly is the first gantry west of Farnborough. You can see the houses backing to Southwood Road on the right and Farnborough Hill in the distant background, the Abbey being obscured by the down working. This gantry was one I could see from my childhood bedroom via a glimpse through the houses on Fleet Road. I lived in the area until 1968 so I'm pretty sure of the points of pedantry though I note this does not appear to tie in with the signalling diagram published. But the South Western Circle, I have just discovered, in their monograph on the LSWR low pressure system does quote two full width automatic gantries west of Farnborough at 1647 yards and 2 miles 47 yards. So I think I rest my case.

'Pedantry point two – On the Kemp Town civil defence exercise on P's 42-3, a caption refers to those wearing boaters being from Roedean school. Well the group looks very male and Roedean was of course exclusively girls. So likely to be from Brighton College, then exclusively boys. I lived in Brighton too in the 70's but Kemp Town had finally closed (though buildings were still extant) by the time I moved there.'

From **Mike Andrews**. 'I always enjoy Southern Times and the above paged reminded me that the Manor Road Footbridge at Farncombe was where I first started trainspotting in the 1950s. This was always known as the 'Iron Bridge' locally. The trains were always EMUs of 4COR, 4RES, 2BIL and 2HAL classes so the late afternoon goods was looked forward to. The engine was usually a Class U or 700 from Guildford shed. Luckily there was a nearby fish and chip shop. Happy days.'

From **Stuart Hicks re ST7.** The caption to the lower image on page 62 is incorrect as the train shown is crossing Hungerford Bridge to arrive at platform 1 at Charing Cross. It presumably is a service from Dartford via Sidcup and Lewisham.'

Finally, and I think this is also rather appropriate as the last letter in this issue, from **Martin James.** 'Please may I make the following corrections to the Staines-Wokingham Part 2 article ...Page 27 Rusham Crossing Signal Box. I worked in Rusham Box as a teenager - blocking trains on whilst the signalman sat outside in a deckchair !!! This block box was always in daily use as a block post. It was manned by two shifts - 6am to 2pm and 2pm to 10pm. It was closed at night - as was the crossing closed to road traffic at night.

'Virginia Water Coal Sidings. The track for these sidings is still in situ buried in the undergrowth - except where the new electrical substation has been built.

'Page 28. Since the building of Longcross Garden Village, for some years now Longcross has had a half-hourly service in both directions mornings and afternoons. Longcross Station has also been refurbished in the last two years - with the exception of the Exmouth Junction Concrete Works footbridge.

'And finally ... you mention AI in your editorial - please may I pass on a quote about AI. AI is like a small child. You don't know what it will do next. You also don't know whether it will turn out to be good or to be evil.'

A plea from the Editor. We do welcome correspondence although on occasions we do have to cut things short slightly. Please could writers identify which issue of ST they are referring to – it really does make life a lot simpler at this end!

On a sad note, some readers may have heard of the passing of Hugh Davies who ran the 'Photos from Fifties' service for many years. It was a facility I used many time. Hugh had run the business from home but also attending shows close to his native Railway Enthusiasts Club at Farnborough. He was a delightful person and will be sorely missed.

Fortunately the collection has been saved and is now operated by Transport Treasury.

Seventy plus years ago was the time of the main line diesels on the Southern Region. Noisy, smelly and prone to deposit oil on anything and everything it would have been hard to visualise how the development of the type would lead to the motive power fleet we have today. To be fair, when they and their Southern sisters were on form they produced a steady performance unlike anything previously seen. We wish the group who are attempting to recreate one of the LMS twins to be enjoyed by a new generation every success.